Government and the Mind

GOVERNMENT
AND THE MIND

JOSEPH TUSSMAN

New York
OXFORD UNIVERSITY PRESS
1977

Library of Congress Cataloging in Publication Data

Tussman, Joseph.
Government and the mind.

Includes bibliographical references .
1. Liberty of speech. I. Title.
JC591.T86   323.44'3   77-74796
ISBN 0-19-502230-0

Printed in the United States of America

Preface

This has been a surprisingly difficult book to write. The bare assertion that government has a serious and legitimate concern with the mind of the citizen, that it is not simply an illicit intruder in the mental domain, evokes such immediate and powerful hostility that I have become a bit gun shy. And since that assertion is only the opening move in an attempt to explore the range of government's concern with minds, I have not wanted to alienate the reader hopelessly in the very act of inviting him to join me in the journey. I am, as a consequence, ambivalent about exploiting the startling quality of an assertion that runs so counter to our received wisdom and that glowers so threateningly. No sooner do I proffer the assertion than I seem forced to defend myself against the obvious charges and to insist that I am not against democracy or in favor of brainwashing or against freedom of speech or hostile to academic freedom, and so on. I find myself, in a Sophoclean phrase, in the midst of a storm of spears—dodging, fending-off, avoiding. Instead of a dashing, joyous, polemical ride I fall into a rather uncongenial, apologetic, placatory mood, indulging only occasionally in a brief defiant gesture. In short, I become constrained, cautious, responsible, almost soothing, where I would prefer, could I afford the luxury, to ride a bit more roughshod.

One result of this initial rhetorical tension is that I pass rather quickly over a range of interesting and complex questions—lightly, not exhaustively, suggestively rather than conclusively, content with a bare sketch and a promise to myself to return to this or that question on another occasion. But the sketch is there. The argument stands on its own feet, though scarcely buttressed. It is, I think, basically unanswerable, but I expect that it may not seem so to everyone. At any rate, with a nagging sense of guilt about the offhand diffidence of the opening section, I move to the consideration of the necessity and legitimacy of governmental action in the various provinces of the domain of mind.

"The Politics of Cognition" is an attempt to bring a number of familiar institutions—the mass media, the artist, the research university—within the focus of a concern for the condition of immediate awareness or social consciousness as well as a concern for the direction of our cognitive energies deployed systematically in the pursuit of knowledge. This is, I think, the most loosely knit section of the book. There is, at every point, too much to say, too much that is both familiar and controversial. Still, it is something at least to place questions of the quality of awareness within their political context and, for the understanding of the life of research, to substitute for the timeworn marketplace model the conception of the cognitive city.

In the section on "The Teaching Power" I make explicit the existence of a tutelary power as an inherent constitutional power of the state, and I pursue the implications of that position through a range of questions of political and educational theory. This is perhaps the most carefully wrought section of the book.

Finally, I try in "Government and the Forum" to step back from the hopelessly complex (and virtually unintelligible) case-by-case development of free speech doctrine in order to present a coherent analysis of government's role in both providing and protecting the forum and in intervening within it. This is a fairly novel and, I think, a badly needed look at the elements of free speech.

By the time we have gone through all this, the original "challenge" my seem remote and even unreal. Of course, government is legitimately involved with the mind! It is so obvious that the original fuss seems pointless. The question is, rather, why I have not explored in greater detail and depth the mass of intriguing questions upon which I touch in what seems always to be "in passing."

The answer is that the problems are really too important and too complex to attempt more, initially, than to marshal them in an illuminating array, to see how they are related to one another, to map the unfamiliar terrain of the domain of the mind seen from the perspective of government. I do have, and even express, positions on many of the problems; but I have worried that indulging my views too freely would divert attention from the essential structure of the argument to the contingent features of my view of this or that controversial question. This fear of imperiling the reception of the central theses about government and the mind by the taint of my transient partisanship led me to toy with the idea of a two-layered book: a central argument—cool, detached, unpolemical, mild, impersonal; and an extended series of appendices or notes in which I could quarrel and abandon judiciousness for free-wheeling partisanship. This idea survives in an impure form. The central argument is intelligible without the notes, but I suspect that some warmth may have crept in. And the notes are not simply polemical asides. They amplify or clarify or expand upon the central text. I believe much would be lost in not reading the notes. But, for a number of reasons, I have not tried to incorporate those that remain into the central argument.

With all the unresolved tension between suggestiveness and exhaustiveness, between judiciousness and partisanship, I believe, nevertheless, that the book as it stands will take us through the mini-Copernican shift in perspective needed if we are to make sense out of the undeniable and ever-deepening involvement of government with the life of the mind.

Berkeley   J.T.
March 1977

Contents

Government and the Mind

1

Government and the Mind

Government, we are told—and by the Supreme Court itself—may not invade "the sphere of intellect and spirit which it is the purpose of the First Amendment to our Constitution to reserve from all official control." It would be difficult to find another statement so plausible, so seductively obvious, and yet so utterly, so foolishly, so deeply mistaken.

That government must leave all that—mind, soul, spirit—alone, that there is a realm free from its authority, another country, a private or a sacred sanctuary beyond the reach of bureaucrat or inquisitor—who would not thrill to that assurance and claim it as his right? What else is the final tyranny than the intrusion into the sphere of intellect and spirit, into the realm, as I shall call it, of the mind?

Nevertheless, I intend to oppose this mistaken principle and to establish not only that government has authority in the realm of mind, but also that its responsibilities there are among the most important that it has.

For the benefit of troublemakers and purists let me state at once that I do not use "mind" in any complicated, narrow, or technical sense, and I do not intend to get very metaphysical about it. There is a perfectly good vernacular mind-body distinction and I will use it freely. I do not pursue the ancient argument that if anything exists it is sort of solid, and that mind is therefore really body or is non-existent. Nor do I limit mind to the purely cognitive, taken as excluding feelings, attitudes, desires. Nor do I make anything turn on distinctions between mind, spirit, and soul. The problems I deal with would not be exorcised by the weaving of a careful web of distinctions. Nor would useful clarity be achieved if I began with definitions. I am attacking the view that government has no business in the realm of "intellect and spirit," or soul, or psyche, or, in short, mind. It is all clear enough in context.

Before taking up the argument in positive terms I will touch briefly on some of the theoretical and historical factors that have contributed to the pervasive American view that government has no legitimate business with the mind.

The persistent conflict in the West between church and state, culminating in a troubled doctrine of "separation," encourages the view that, like amphibians or commuters, we are involved in two distinct realms—a heavenly city and an earthly city, a city of the spirit and a city of the flesh. The secular state is, as a usual consequence, thought of as having a severely limited jurisdiction. It is not to interfere or meddle in the affairs of the spirit. That is the business of the church, or the churches, or the private conscience. Caesar may have what is his, but he may not touch what belongs to God or to his feeble heir. Thus, our disposition to regard mind, in a broad sense, as beyond the authority of government grows, in part, out of the sense that the life of the spirit falls under a religious jurisdiction.

Philosophy, apart from religion, contributes also. Mind and matter, thought and extension, consciousness and lumpishness, ideas and things, mental and physical, are some of its familiar dualities. With such distinctions at hand a "person" is easily, if crudely, thought of as a mysteriously bonded two-layered entity or perhaps as a kind of sphere with an external shell and an inner core. Usually, the shell is physical and the core mental. The mind is inside; the mental life is the inner life. Inner and outer coincide in some contexts with private and public. Words and gestures are public; what I think, what I really think, offers itself as the clear case of what is private. So, as private, the mind lies beyond the reach or jurisdiction of public authority. Government may take notice only of the outer persona, of acts, not of thoughts. Thus, the mind-body distinction (necessary as it is) is brought to bear upon the delineation of the scope of public authority and contributes the privacy of the mind to our stock of basic notions. Public authority is unlicensed in the private world. It must leave the mind alone.

This conviction is reinforced by a popular conception of sovereignty as "the supreme coercive power." Government action, on this view, is essentially imperative and coercive, commands and sanctions, orders backed by threats. Government is seen as the policeman pushing people around at the direction of rulers who lay down the law. It is organized force. In a nutshell, government is "the law," and "the law" is the cop on the beat. Its ultimate symbol is the club.

Seeing government this way we see also that it is unqualified as an instrument of our higher aspirations. Clubs are for skulls, not for minds; for bodies, not for souls. You cannot command virtue or order up morality. We are to be clubbed only for what we do, not for what we think. The police, the law, government, should leave the inner life, the mind, alone.

A popular folk-constitutionalism is also something of a culprit. Important as are the United States Constitution, the Bill of Rights, and the First Amendment, they tend, unfortunately, to generate some misconceptions. They encourage the unwary to think that we are denying constitutional authority to all government when we may, in fact, only be denying some authority to the Federal Government. For example, the Establishment Clause of the First Amendment denies to Congress any authority to make laws "respecting an establish-

ment of religion." But its adoption was not regarded as preventing the States from establishing or dis-establishing religions, as they saw fit. The First Amendment as a whole must,
historically, be given a "states' rights" reading; it denies to
the newly created central government powers over "the
sphere of intellect and spirit" jealously reserved to the states.
The Fourteenth Amendment subsequently affects the situation, but it is a serious oversimplification to say that it places
beyond the reach of all government, federal or state, what was
originally placed beyond the reach of the Federal Government alone. This is a widespread misconception which plays
havoc with attempts to make sense out of the Constitution as
it bears on the mind.

And, finally, distrust of government in general applies with
special force to the domain of mind and spirit. Whether we
think of the depressing tendencies of large-scale bureaucracy
or of the strident electoral and legislative processes, we are
not comforted by the thought that subtle, precious aspects of
life may be brought under that sort of rule. The still-living
memories and the continuing burden of modern dictatorships with all their Orwellian spiritual horrors foreclose any
expectation that the inadequacies of bureaucratic democracy
can be cured by dictatorship. Government seems always at
least on the verge of becoming an evil; and we seek, as did
John Stuart Mill, a protective principle which would define
and limit its scope no matter how benign its intentions.[1] That
government has no authority in the realm of mind seems to
be one such safe and valid principle. Carlyle's phrase, "anarchy plus a constable," becomes, in our version, *"spiritual*
anarchy plus some law and order."

The weight of experience and reflection which thus expresses itself is quite formidable, but I comment here only
briefly.

Something important is expressed by the distinctions between church and state, religion and politics; but "separation" is a peculiarly misleading way of talking about deeply interrelated aspects of a single thing. A single community may have distinct political and religious organizational structures; a single person may have political and religious commitments. Interaction and conflict are almost inevitable. Nothing is aided by the imagery of compartments, of "walls of separation," of independent areas.[2] Historically, the relations between churches and governments have taken many forms, no one of which can claim a monopoly of legitimacy. Even dis-establishment does not avoid, as our own experience has shown, problems of mutual support, cooperation, friction, opposition. Political decisions about secular matters may affect religious life without, on that account, being subject to religious veto; nor does religious motivation necessarily override the barriers of the law. Not even the greatest respect for religion, especially where concern for religious freedom insists on dis-establishment and pluralism, can confine the authority of government to that which does not impinge on the soul or invade the sphere of the spirit.

Secular versions of the mind-body distinction are also useful in a variety of contexts. Nor do I have any serious quarrel with "inner" and "outer"; although, of course, they are spatial metaphors which do not really illuminate the relations of mind and body. Nor would I cavil at some identification of the "inner" with the "private" and the "outer" with the "public." Provided that we can resist the temptation to pun. "Public" as the outward show, the public face, is one thing. "Public" as broadly consequential (as in "affected with a public interest"), as a matter of general concern and therefore subject to public authority, is quite another. "Private," in one sense, may be the inner or the mental; in the other sense, it is the generally inconsequential. Obviously, they are not to be confused. What we think, feel, and believe may be "private" inner states, but they spill over into conduct that affects others. The condition of mind about obedience to law, for example, is at least as determinative (more!) of the stability of a social order as is the policeman's club. The mind may be inner or private, but its condition is clearly a matter of deep public consequence, of legitimate public concern and, unless something more is added to prevent it, within the scope of public authority. In short, the sense in which the mind is private is not the sense in which it is not subject to public authority.[3]

The tendency, when we think of government, to think simply of coercion, of commands and sanctions, of law-maker, judge and police, is simply a failure of understanding and imagination. Consider, for example, the public school. It is a governmental institution as clearly as is the fire department, the board of public health, the municipal court. The governing structure of a state university is as much a part of government as is the city council of the community in which the university is located. The schoolteacher works for the government as unmistakably as does the deliverer of the daily mail. In fact, if we consider the question afresh, we may well conclude that the public-school teacher in America today is the most appropriate symbol of government in action, the paradigmatic government agent. Government acts in a variety of modes and it is not precluded, simply by virtue of a narrow misconception of government as essentially coercive, from acting deliberately and appropriately on the mind.[4] There is, of course, a coercive aspect of government, and there is something to be said for the view that government ought to enjoy, if not monopoly, at least coercive supremacy. But to say this is not to say that the essence of government is coercion or that it can only act in this mode. A parent may also lay down rules, warn, and punish, but that does not define the essence of parenthood or the limits of the parental role.

As for the general misarchy or archophobia which, out of a proper detestation of tyranny, seeks to limit government at every possible point and, therefore, at the edge of the mind—what more can be adduced than the traditional insight? Anarchy is not the answer to tyranny; it is its breeding ground. The real enemy of tyranny is that great creation of the civilized mind, legitimate government. And it is, precisely, with the elements of legitimacy in the relations of government to the mind that I am here concerned.

But why so grim, so timid, so apologetic? Why this fanfare and drum roll for a dogma that should be quietly buried? It is some sort of tribute to our scars and our fears, some showing of respect for a deep mistake in a good cause. Enough! There are perfectly good, straightforward arguments which place the legitimacy of government's concern with the mind in a natural and benign light. Let us look.

A community is constituted by—its very existence depends upon—a condition or state of mind. It is not a mere collection of physical entities or a herd of biological organisms. It is a continuing organization of persons related by shared understandings, commitments, agreements, attitudes.

There are, of course, reductionist tendencies at work as we attempt to understand the community or polity, and it is always tempting to settle for something more tangible and more easily observable than a state of mind. Heads and hands can be counted, the movements of herds can be mapped, the play of power traced and measured. But we can never, in these terms, tell the central human story—the story of conscious human beings creating and nurturing common enterprises and fellowships in pursuit of shared visions, struggling for some semblance of peace, freedom, justice, dignity, brotherhood. To try to understand a community without attempting to grasp the condition of mind that distinctively constitutes it is to systematically miss the point. Not simply because a community is the sort of entity that happens, also, to "have" a mind; but, more fundamentally, because community *is* a condition of mind.[5] If the constitutive condition of mind is lost or absent, there simply is no community in any significant sense.

A related way of putting the same point is that man is a political animal—an animal transformed into a person by the operation of and involvement in a community. As expressed in a great Platonic allegory, man is a marsupial and the community is the nourishing second womb within which he continues development toward his real birth as a person and member. There, in the limbo of minority, he incorporates the language, awareness, attitudes, and understandings—the mind—without which there is neither community nor person.

But if we are uneasy about these abstractions and images, let us remind ourselves, more prosaically, of the general dependence of undertakings upon understandings.[6] Consider some actions: the moving of a piece of ivory on a board, the exchange of a piece of paper for some food, the marking of a ballot. The physical act is simple; but before it can be a move in a game of chess, an economic transaction, or voting in an election, the physical act must take place within a complex context of understandings. It is the mental context which is crucial. If it is absent or distorted, there is merely a pointless physical act or, at most, a parody. To create and sustain a game, a monetary system, a political order, is to create and sustain the appropriate state of mind. To be concerned with undertakings is to be involved with minds.

These assertions need, no doubt, more extended analysis and defense, but I will take it that, benignly considered, the central point is clear enough. Any sustained human enterprise requires and rests upon its appropriate condition of mind. A polity must, if it is to continue, recruit and incorporate new members; it must provide for a fruitful life of communication; it must direct attention to its problems and cultivate the knowledge and wisdom it needs. Or it will die. There is, therefore, an overwhelming public interest in the condition of the mind, that, without exaggeration, may be regarded as the most fundamental part of the public domain.

The community, being what it is, has a great stake in the condition of its knowledge-creating and transmitting institutions, in its institutions for informing, discussing, deliberating, deciding. The question, then, is whether government, the instrument through which the community acts to promote so many of its aims, has (or may be given) authority to act in this area of concern, or whether, for special reasons, the community may not act through government in matters which concern the mind. I am not, at this point, raising questions of the desirability or wisdom of this or that mode of governmental action or of whether it would be wiser to leave the field entirely to non-governmental institutions. The preliminary question is, rather, about the legitimacy of a community's assigning to its government authority to act in the realm of mind, to invade "the sphere of intellect and spirit."

The weight of the world is on one side of this question. Government everywhere—revolutionary or traditional—is deeply involved in education, in the governing of communication, in the direction of cognitive energies. It is unlikely that a "consensus gentium" would be mistaken about so fundamental a question of authority. The point is not that "de facto" makes "de jure"; there is, rather, agreement about what is "de jure." The United States is no exception. Consider its vast public-school system. It may have problems and failings, but does its very existence violate the Constitution—even the First Amendment—or the higher law? Is the Federal Communications Commission illegitimate? Public univer-

sities, laboratories, research grants—all illegitimate? The
weight of anomalies crushes a purported principle that government has no authority in the realm of mind, or the parochial view that at least the First Amendment—*our* crowning
glory—stands as a barrier between government and the mind.
The school board is, in principle, legitimate.[7]

If I seem unduly assertive it may be because I am baffled at
how to demonstrate the obvious. Questions of legitimacy get
more difficult as they move out from relatively clear contexts.
A particular agent or agency of government acts. Has it been
authorized to do so? Can it show its warrant? Has due process
been observed? Yes. Has some relevant rule or constitutional
injunction been violated? No. Then the act, wise or not, is
legitimate or, as we say in this context, constitutional—within
the bounds of the agreed-to authorization. But then, we may
ask, what about the Constitution itself? Does it observe the
rules of legitimate constitutionalism? Does it violate the rules
of morality or justice, or the higher law, or the will of God,
and thus, in spite of local sanction, lose its claim to legitimacy?

I certainly would not limit questions of legitimacy to questions of constitutionality. Powerful as the notion of agreement may be, agreements are subject to legitimacy criteria.
And so with constitutional provisions, even if constitutions
are seen as constitutive agreements. But where is the principle of morality or justice or the higher law that holds that
a community may not assign to its political institutions, to
government, responsibility for various aspects of the life of
the mind? There is none. A community may, if it sees fit, bar
its government from acting in this or that area (although
even here there are questions of inherent or necessary and
inalienable powers) and it is interesting to study the Constitution and the First Amendment in this light. If one emerges
from such a study with the conclusion that, in America, gov-

ernment, federal or state, is generally barred from the sphere of intellect and spirit, then he had better go back to the drawing board and start all over.

The upshot is that the dogma that government has no business in the realm of mind is unfounded and radically false.[8] Its unthinking acceptance hinders our understanding of the difference between legitimate and illegitimate action *within* the field. Why should we condemn ourselves to live with a shallowly seductive principle and a herd of anomalies when the alternative makes so much more sense. Government has (to a degree which we will explore) authority over the mind. Recognizing that, we must develop the categories and concepts that will protect us against abuse and bring more wisdom into our attempts to deal with some of the most crucial of our problems.

That is what I now propose to do. First, I will discuss problems of awareness and of the direction of cognitive energies—what I think of as the politics of cognition. Second, I will consider education as the expression of an inherent power of the state—the "teaching power." Third, I will discuss the proper involvement of government in the life of the forum—the government of communication. Under these headings I cover, I believe, the most important ways in which government is involved directly with the mind. There is, of course, some overlapping and much interdependence, but that is unavoidable and harmless.

But wait! One last feeble defensive twitch. I am so accustomed to being charged, long before this point, with not understanding the uniqueness of Democracy, of being a crypto- (or not so crypto) totalitarian, of favoring the replacement of freedom of mind by systematic brain-washing, that I must indulge myself in a few anticipatory disclaimers.

No, the democratic conception of society or government is

not at odds with the view that government is legitimately concerned with minds. The democratic character is, in some respects, distinctive. It is very complex, involving the habits, attitudes, and understandings appropriate for participation in political life. It is a product of deliberate nurture. The democratic way of life is not a condition men fall into if only the mind is left to its own devices; it is not the fruit of spiritual anarchy or neglect. However much democracy may differ from other forms of polity it is not, in principle or in practice, an exception to the rule that the community may legitimately act, through government, on the mind.[9] The burdens and responsibilities are merely greater.

No, there is nothing inherently totalitarian in a position which considers that the operation of a public educational system, the support of science, the arts, and institutions of communication, may be legitimately assigned to government. There is, as we know, a tendency toward governmental intrusion, monopoly, and centralization abroad in the world. That is the genuine totalitarian tendency. I do not encourage or support it. My own inclinations are strongly pluralistic. My general thesis leaves questions of the desirable degrees of centralization or decentralization, of modes of governmental or private control quite open, as important questions of policy.

No, I do not propose that we substitute the brain-washed for the free mind.[10] Of course, if we think that providing schools and things of that sort destroys the freedom of mind, then we may be in trouble. It certainly is possible to corrupt and weaken minds, to destroy imagination, to promulgate shallow dogmas, to barbarize the life of communication, to foster ignorance. It is even conceivable that all this should be done as a deliberate policy. I fail to see, however, why this should be regarded as an inherent consequence of the doctrine that government is legitimately concerned with the mind—any more than saying that government is legitimately concerned with public health implies that it must therefore be committed to cultivating a variety of diseases.

No, I am not trying to undermine democracy, to promote totalitarianism, or to destroy intellectual or spiritual freedom. I simply seek to bring to the support of our practice, in a crucial area of our political life, a more adequate theoretical understanding, to lighten a burden of guilt and incoherence.[11]

2

The Politics of Cognition

The history of speculation about the connection between politics and knowledge goes back, at least, to Plato's conception of the philosopher-king. The art of governing, he thought, had something to do with knowing what ought, or needed, to be done. This was and is something of a novelty—almost a joke—since ambition, hunger for power, the habit of command, luck, wealth, sympathy, partisan intensity, a diffuse pandering proclivity, and, of course, charisma are so obviously the more reputable bases of claims to the scepter. While I sometimes find myself, if I nod, sympathizing with Plato, I will not pursue the theme of knowledge as the basis of a sound politics, but rather will be concerned with the way government may, should, or must intervene in the life of cognition—in the cultivation of awareness, in the sustained pursuit of knowledge, in the restraining, even, perhaps, of unholy, shameless curiosity. I will treat the condition of awareness as a political question, as I will also treat, in turn, the direction of cognitive energies, of research, of science. If I had to select some pieties to flout, they might be, for "awareness": that public awareness or public opinion, however it generates itself, is to be served by government, not interfered with; and for "research": that knowledge is always a good thing and that its autonomous pursuit by its devotees is not to be impeded or restrained. So first:

The Politics of Awareness

We do not, as individuals, entirely control our own awareness. We are creatures of immediacy troubled by deep-seated
and transient desires. We are in the grip of urgencies, buffeted by impressions, sensations, pressures, and exposed to in-
cessant importunings. But we also develop goals, make plans,
learn to ignore distraction and to some degree and somewhat
sporadically focus our attention and put our minds to problems of our own choosing, according to our own agenda, regardless, even, of mood or inclination. We transcend immediacy in our grasp of concepts, generalizations, patterns,
plans. We discipline ourselves, try to take charge, to steer
rather than drift. Still, what is on our mind at any moment is
not altogether of our own choosing. Concentration is difficult.
The mind wanders. The mood changes. We lose our sense of
direction and drift again. What we are aware of is not necessarily what we want to be aware of nor what, in some complicated sense, we should be aware of.

We may speak of public awareness, what a community is
aware of, in comparable terms without, I think, conjuring up
and having to exorcise metaphysical ghosts. I am aware, we
are aware, they are aware, everyone is aware—perfectly good
locutions, some riskier than others, and nothing that we
should boggle at. Public awareness is, like private awareness,
the creature of many forces. It reflects needs. Its focus and attention wander. It lives on crises and excitement. It is fickle
and evanescent. It has its own rhythm of seasonal preoccu-
pation.

Considered instrumentally, as playing a part in our lives,
awareness has its failings. We may, at times, be inattentive,
unperceptive, heedless, preoccupied, absent-minded, distracted, diverted, forgetful, slow to notice, to "see," to realize,

to appreciate the significance of, to grasp what is before us. There are failures of dullness, of seeing too little, too slowly; and there are failures of sharpness and quickness, swamping us with detail and irrelevancies. We may not be aware of what we should be, and we may even, at times, be aware of too much and of what we should not be aware of at all.

Failure in awareness is not as simple or clear-cut as failure in belief where the question is merely whether a belief is true or whether an inference is valid.[12] It is not raining and I leave the house without my raincoat. I did not notice the clouds moving in. It is not that I judged, falsely, that it would not rain. I failed to make an obvious inference from what I should have noticed. I was distracted or preoccupied, or absent-minded—a failure, I would agree, of awareness. By "awareness failure" I have in mind a failure of the machinery of perception, a misdirection of attention, failure in noticing, inadequacy of focus, systematic distortion, perspective failures—that sort of thing.

Awareness—or, perhaps, attention—is selective. The mind does not simply reproduce in consciousness whatever is out there; it is not a mirror or a photographic film, a conveniently portable copy of the whole world or even of that part of the world contiguous with our senses. If one holds some such view of the mind as a passive recipient of all the signals beamed at it, one must still distinguish between what is, in that sense, in the mind from what is on one's mind, or at the front of one's mind, or of what one is aware of or paying attention to. On any view, that is less than everything.

Paying attention is a kind of focusing. Our attention is caught, directed, brought to bear on something—here or there, past or present or future. Our focus may be broad or narrow, sharp or blurred. When we are thinking about something we are ignoring what is out of focus. And within our focus there are problems of notice. We may seem to be watching the same game, but one of us notices the strategy, another the footwork. The expert sees what the tyro does not. Noticing is a function of knowledge, experience, interest, and character. A partisan may clutch at straws only a partisan eye can even discern. An honest man may be blind to signs of chicanery, an obsessive may notice nothing else. The optimist notices improvement where the pessimist sees only failure. Dr. Watson probably can count, but he simply walks up the stairs; a freak like Sherlock Holmes notices that there are seventeen (or something!). What "is there" may escape notice, or be beneath it.

Awareness is crucial to the way we conduct our lives and it is important to realize that flawed awareness may betray us. We can be blind or obsessed, lulled into indifference to real dangers, diverted from real problems, stimulated into hysteria over trivialities. Consciousness, individual or social, is not, simply by virtue of what it is, always what it should be. It may be inadequate to our needs. It may even be diseased.

The question, then, is whether government has a legitimate role to play in the shaping and protecting of awareness. There is, of course, the orthodox view—the standard error—that government should do nothing about public awareness except respond to it as quickly and as faithfully as possible. This is offset by the habit of heaping blame on government for having failed to warn, to awaken, to force upon public attention what it ought to have been aware of. On the one hand, no manipulation; on the other hand, leadership that enlightens, teaches, and forces us to attend to the necessary agenda.

A modern dictatorship may attempt to take complete charge of consciousness. It can monopolize education and the means of communication in the service of its own awareness policies. It can combine "bread and circuses," minimal-want satisfaction plus distraction, with militant mobilization against the enemy of the day—other states, or inequality, or sloth, or selfishness, or the remnants of outmoded consciousness. It may promulgate its creed and protect it from challenge. This may sound rather drastic, but there are still bitter and mischievous intellectuals who, in childish despair over the obdurate consciousness of the average man, believe that nothing short of a sustained and drastic tutelary dictatorship can transform the poor wretch into the semblance of the new man fit to enter the promised land. To this "revolutionary" mind nothing, in the shaping of awareness, is forbidden to the dictatorship of the chosen; and, of course, nothing is to be permitted to a democracy attempting, diffidently, to protect itself against grosser forms of ugliness, folly, or insanity.

Far short of guardianship there is, even in democracies, a role for government in the shaping of awareness. A Churchill is honored not only for his attempt to warn of things to come, but for his capacity to make the country see itself as involved in a meaningful struggle, to lift its spirit in adversity, to sustain its awareness of its destined task. Apart from war, our great political heroes are often awakeners and summoners, creating the appropriate mood, expressing the necessary idea. And even routinely we expect government to direct attention to problems, to make us think about energy or pollution or population or crime. In various ways, legislators, executives, judges try to shape and reshape public awareness. Without claiming monopoly, government must, at least, enter the struggle for attention.

It does so through its involvement in the special institutions of awareness. Since I will be discussing the school in some detail I defer educational questions until later. I defer also, for later discussion, the general problems of government in the forum. But some discussion of awareness institutions is appropriate here. At the private level we have our checklists, mnemonic devices, calendars and diaries, our rituals of withdrawal, stock-taking, contemplation, encounter, stimulation, and oblivion. Groups or societies have their holidays, commemorations, rituals, symbols, catechisms, monuments. Our lives are hedged by signs, alarms, warnings, and reminders. But traditional devices pale into insignificance in the shadow of the press, the screen, the tape, the tube—the massed media.

The classical distinction between appearance and reality, between the way things seem and the way they are, so central to problems of knowledge, has some difficulty in accommodating itself to the world of the theater. The proscenium arch marked the point of entry into a different world; a play was make-believe. "Once upon a time" was a special clue; we were to suspend disbelief and attend in a special way. The stage was not continuous with the ordinary world. We were presented with a rehearsed fabrication. There was a contrast, unconcealed, between show-business and reality.

The sense of something staged comes more and more to pervade the public world. The news that edifies us in the morning is an editor's composition. The television screen brings us a news-show, presided over by that new Virgil, the trusted anchor man—filmed, taped, trimmed even when live and in living color. There is the arena, the cast of characters, the commentator-chorus. The public stage has almost ceased to be a metaphoric expression. The public figure is an actor worried about his rating. Communication technology has presented us with a pervasive screen, with images and sound, upon which, when we are aware of public affairs, our attention is fixed. Tuned in, we live in an edited world. Somewhere between appearance and reality is the strange world of videality.[13]

What is most significant about videal appearance, about what appears on the screen, is that whatever we see has been seen first by a camera and has survived cutting. It is a selection of a selection that comes to us, impressively (one picture is worth a thousand words!), as the world, or as what is happening in the world. It has already been digested when it is offered for our nourishment. Our attention is directed to what has already succeeded in capturing attention—someone *else's* attention; and while much has been screened, infinitely more has been screened out. We are, in an edited world, at the mercy of the editor. Who, then, writes the scripts, points the cameras, produces the big show on which, for an astonishing proportion of our time, our eyes are fixed? Who thus directs our awareness? Who should?

One answer, as I have already suggested, is that government should seize the camera and projector and offer to its subjects the edition of the world which would best serve the needs of the governed, or perhaps of the governors. Why should we not see what confirms the wisdom of the policy on which we are launched? Why encourage malcontents by focusing on temporary difficulties? Why not prepare people for the necessary sacrifices by emphasizing the evils to be combated? Why not whip up some healthy indignation? Why not . . . ? Monopolized and carefully directed, the mass media can be interposed as a politic curtain between the public and the real world—the unabridged, unexpurgated version.

One would have to have more faith in the wisdom, the virtue, and the competence of government than seems, at this stage of the world, reasonable, in order to endorse monopolistic governmental control of the mass media. But the rejection of governmental monopoly does not imply that government can or should completely wash its hands of the problem. We may not—I do not, at any rate—want to live within a government's version of reality. But if, as it seems, we are doomed to live with mass media, we are, perforce, in an edited world. We may reject the government edition, but we cannot escape the editor. And who, to repeat, is he; and by what right does he shape our awareness? And may government have a role in protecting us from him? Must it not, in fact, develop and enforce some essential elements of awareness policy?

Two models come quickly to mind: the awareness "business" and the awareness "profession." Considered as a business, the traditional wisdom about government and business can be brought to bear on the problem—some anti-trust stuff, some consumer protection, some licensing and public-policy restrictions, etc. Nourishment for the mind is just another commodity. And, just as in other areas, believers in the marketplace as the best of regulators will argue that a free market in awareness and ideas maximizes something—wisdom, or understanding. It is an interesting position, and, as is so often the case, the best argument for it seems to be the suggested alternatives. I have a deep-seated reluctance to bring the life of mind and spirit under the aegis of the marketplace, but I find myself softening around the edges as I become skeptical about the depth of the wisdom we bring to the regulation of our affairs. The market at least has variety, something for everyone and for almost every taste. It is less concerned to alter human nature than to accommodate itself to it, more deferential than reforming, more engagingly vulgar than censorious. It has its familiar vices as well, but, all in all, we have seen worse rulers than the marketplace in this century. If one is concerned, practically, to limit the power of government one must recognize that nothing can stand its own ground so effectively as institutions rooted in relatively independent economic power. So, if we fear big government, we must hope for business big enough to defy it. When we listen to a commercial and learn that we are indebted to some powerful corporation, we should quell our annoyance and consider the real alternatives, and perhaps, even, appreciate our debt.

But while I certainly reject "public ownership of the means of consciousness production" and am prepared to come to some terms with the marketplace of ideas and the awareness business, I insist that we should not deal government out of the game. It has a role to play, which I shall consider more fully in my later discussion of government and the forum.

In contrast to the idea of awareness as a commodity to be marketed stands the "professional" notion of proper awareness as a condition to be achieved through the services of the awareness professions—journalism, for example. The professional conception is that of a corps of men and women entrusted with responsibility for some great social value—as justice, or health—who develop the skills, the traditions, the commitment, the morale necessary for the guardianship of what has been put in its trust. The profession is dedicated to its mission. It is, presumably, trustworthy and has procedures for purging itself of its rogues. It claims a large measure of autonomy and is quick to protest interference—amateur, or political, or even economic—as impeding the performance of the function for which it is responsible. The profession may, out of necessity, learn to live within a framework provided by church or government or corporation, and in some sense it outlives these transient hosts. The church may crumble, the government may fall, the corporation wax and wane, but the heirs of Aescalapius continue in an unbroken fellowship— the guardians of health. It is a great idea.

Much of what is familiar when we think of medicine and law may be relevant also to the professional component of our massive awareness institutions. There is the corporation, the network, the independent broadcaster—the owner or business interest. There is government circling restlessly around. And there is, let us say, the working press—printed or electronic—whose "professionalism" is the real basis for the claim to freedom from interference.

I will not try to judge the performance of the press; I am beneficiary and victim, but hardly a qualified critic. I will touch, however, on several points related to its professional function that tend to bring into doubt the claims to autonomy and freedom from political intervention. The conceptions of "objectivity," of the faithful reporting of what happened, of the messenger dutifully bringing the message from the world, the good news and bad, of the unbiased transmitting medium—these familiar notions do not seem really to capture a situation which is so essentially one of attention, selection, focus, perspective. A picture (like a soundtrack) assuredly seems to record what happened. But it is really a kind of visual quotation—something taken out of its context. "Objectivity" is not an answer to the relevant questions: Why was the camera there? Why focus on that? Why so oddly out of context? We seem to need, instead of professed "mirroring" objectivity, notions of significance, of interpretive stress and balance, of fairness. And it is not clear that these notions are either understood or embraced by the "profession." Cynicism about a naïve conception of objectivity is about the high point of its intellectual achievement. Joined with a naïvely "realistic" disavowal of concern for significance or fairness ("we're all biased!"), protected by a thick layer of self-righteousness and sanctimony and lo!—investigative journalism. Energetic and restless, it scratches and rummages. If a bone is glimpsed there is a swarm, a stampede, and, in its midst, a general excavation, a cloud of dust and, believe it or not, a "deeper understanding" of our skeleton-burying corruption. It has yet to be grasped that a prolonged and intensely focused gaze is as likely to produce misapprehension as apprehension. But, of course, don't shoot the messenger, or the digger or the scavenger—he's only doing—smirk and pause for congratulations—his job.[14] His not to judge, or interpret (unless he happens to be our commentator); his but to bring you the facts so you can make up your own mind.

And yet, tempting as it may be, let us pause before we rush in to repoint the camera, to redirect attention to the really noteworthy, to restrain the shameless exposure of what is hidden, to elevate our consciousness by cutting down the coverage of crime, violence, sex. It may be that this journalistic élan under the loose curb of commercialism serves us better than we think. Fear of the reporter, striking randomly, may be all that restrains many in power who no longer fear the eye of God. (The dogged reporter as the Hound of Heaven.) And as for where the camera should *really* be pointed—can something not be said for violence, crime, sex? Civilization is a sustained struggle to tame all that, to domesticate the animal, to build on earthquake territory. What is more natural than a fascinated concern with every tremor, with every baring of the fang, with every eruption of the state of nature which it is our perpetual task to subdue? Homer, with an unerring eye for the significant, sang of nothing but strife, violence, fraud, sex. Every crime, every act or threat of violence—lawless or lawful—brings us back to the heart of the matter, to our fundamental problem. It fascinates because it is significant. And what is more significant? Every petulant whine about equality? The shortage of oil? Johnny's failure to read? Zero population growth? All of these and more, brought duly to our attention by the undirected (by government, that is) media?

Our mistake, when we put aside the paper and turn off the news with a sigh, is in taking it for what it is not. It is not, nor is it meant to be, a picture of the world with everything there in proper perspective—joys outweighing sorrows, quiet virtue prevailing over marginal evil, achievement and triumph overshadowing failure. It is a selected budget of disasters or threats, of contests and confrontations, of things which need attention—relieved by whimsy, human interest and occasional reports of victories. And this is, in fact, a *pro-*

fessional product, an edition, which should not be lightly put aside in favor of a system of awareness-shaping administered directly in response to a political determination of the proper awareness agenda.

This is, I think, more defensive than I had expected or intended, but let it stand. The question is whether there is much of a case for the application of the conception of professionalism to the media as a basis for the claim to autonomy. If I begin with some dissatisfaction over the status of conceptions of objectivity and news-worthiness, I end, nevertheless, with considerable respect (grudging, reluctant) for the professional product. Enough respect, at any rate, to restrain any impulse to add "Awareness" to "Health, Education, and Welfare."[15]

There are, however, problems beyond the difficulties of objectivity and disinterested professionalism. The power of the media may be drawing to it many who do not see themselves as professionals disciplined into objectivity but as men of commitment, yearning to awaken society, to enlighten it, to transform its consciousness. "We care," they say, as they prepare to foist their interpretations of the world upon the passive viewer. "We have values, we take sides, we serve our biases openly, we are honest moral agents, not hypocritical neutrals." It is as if the referee had decided to take a hand in the game.

That this credo of the new breed would, if it prevails, undermine the claim to special status, access, and privilege essential to the activity of an autonomous media may be only dimly perceived. The autonomy of the profession can be justified only (as is also the case with teaching) if we can make sense out of the distinction between what the role requires and the private desires, partisanships, and private commitments of the profession's members.[16] Any professional must learn to live with that distinction. If it is lost or rejected, the basis for special trust is gone and the natural, inevitable, and utterly justified question will raise its head: Who are they that they should have this position of special influence? Who elected them to foist their private foibles on us? Why should *they* have this disproportionate voice? Why indeed?

A variant of the private commitment theme, also rejecting objectivity, is the view that the media functions primarily as "the opposition"—that the proper position of the media is "at the government's throat." This is, of course, a very American position and would hardly be understood in many still dark places on the globe. It is hard to take seriously the notion of the awareness institutions of a society as primarily in the service of opposition and discontent. On the face of it, it is absurd. At bottom, it is more profoundly absurd.

The argument, if one looks for something besides paranoia, is that government is too powerful and that in spite of separation and distribution of powers, checks and balances, built-in oppositional devices like parties and elections, judicial appeal, and countervailing powers like corporations and unions, something more is needed. The press, the media, appointing itself "fourth branch" assumes a systematic oppositional role. It simply opposes the government as if it were the enemy—a duty of love, of course. It plays gadfly and, in a horrible misconception, even gives itself Socratic airs. To oppose without discrimination? Even if the government happens to be gloriously right, decent, on the side of the angels? Oppose what is good and right? Well, of course not. We oppose what is bad and support what is good—as we see it, of course. So "we"—the owners? the network elite? the working professionals?—decide what to oppose and what to support and deploy our power in support of our policy. We point the camera and edit the videal world to fit our script. We play our politics.

By this devious route the "non-political" awareness institution, the guardian of the integrity of public consciousness, the privileged professional corps anoints itself as a branch of government and sees itself as a political force, a special kind of political partisan deploying its unique weapons—control over the camera and microphone, control over time, a deciding voice in the struggle for attention, to further its politics. In thus politicizing itself, in relinquishing professional neutrality, it invites government to enter more fully into the politics of awareness. Deeply consequential, affected with a public interest, its objectivity or neutrality shaken, the mass media are not placed, by the higher law, beyond the authority of government.

Beyond the technical institutions of communication, more fundamental in the shaping of human awareness, looms the figure of the Artist. Working through word, color, sound, shape, movement, he holds a special place in the story. A Homer forms the mind which acts out the history of Thucydides; a Milton gazes, for us, into the depths of rebellion; a Bach gives us a glimpse of order. Civilized consciousness is parasitic upon special insight. The magical power of the artist, our indebtedness, our mixed feelings of love and awe, of hate and fear—these are the elements of the situation. The genie is out of the bottle and, although he seems to serve us, we worry about controlling him.

What puzzles us is the relation of special insight to ordinary vision. Here is the great, central, standard, ordinary common sense, geared to the world of ordinary life and ordinary action, the ordinary order of things in which we ordinary people ordinarily live, corrected to twenty/twenty. And there is that madman, over-sensitive, over-perceptive, too sharp, too discerning, seeing what is beneath or above or beyond our notice and even showing it to us. Are we too dull? Or are we

just right and they too sharp? Who is tuned in to the real or
right world? Blake or the butcher? The village band or the
Pied Piper? Demos or Plato?

The dull answer, no doubt, after we have quarreled sufficiently, is that the extraordinary stands on the shoulders of
the ordinary and lends it illumination and vision. If it escapes, to stand alone, it becomes inhuman; if it is kept in
service it enhances, shapes, and reshapes our ordinary awareness.

We are prone, in a Romantic tradition, to cast the artist in
the role of rebel as if he is doomed, by his insight, to be
against his society. But that is too limited a view. Art has a
celebratory as well as an oppositional function and the former is at least as important as the latter.

The celebratory function is that of revealing the significance of what, to unaided perception, might seem ordinary,
prosaic, routine, unimportant. The artist takes the familiar
object and shows us that it is to be marveled at; or shows us
that we are playing a part in a story more exciting than we
realized; or shows us that a meal is a profound event. He is
the master of ceremony and he surrounds us with significance.
So, in this role, art builds our cathedrals, adorns the chapel
and the nest, composes our masses, hymns, and anthems, fixes
our hero in the great act, turns toil into a struggle to achieve,
and keeps us whistling while we work. It lends, to the ordinary life, a deep enchantment.

There is the other function too. The same fairy godmother
can, at a stroke, turn the prince into a bumpkin. The discernment which adorns can also strip naked, make pomp pomp-
ous, turn the nose into a demeaning joke, destroy with caricature and parody. Art can unmask and disillusion; it can expose
the fraudulent. It can also make us sneer at what is sacred and
deride greatness. Vision is remorseless and hard to tame. Here
is Aeschylus celebrating Athens and the Law, the taming of

the Furies; and there is Aristophanes leading a great, prolonged Bronx cheer from the gallery. Art serves both order and disorder, establishment and rebel. It helps us both to see and to see differently.

I do not mean to suggest that the artist sees it this way; that he is concerned, necessarily, with his "function"; that he cares to serve anyone. He may care only about his art, his vision. But his work has consequences in awareness, and at least since Plato regretfully escorted the unruly poet from the city men have sought to control his power, to harness it to their purposes, to bring it and keep it under law. This effort is a long chapter in the story of government and the mind. I will touch here on two chief modes of control—control of the supportive institutional structure of the arts and direct control of substance or content.

The artist needs support for his way of life. He needs time; he needs materials; he needs access to his special kind of audience; he may even need a supportive social milieu, colleagues, worshippers, critics. Some system of patronage must provide resources and opportunities. Sometimes it is provided officially, accredited status in the official guild being a necessary condition for the open practice of an art. In some systems, public patronage may be selectively bestowed and not tied to license to practice. Sometimes the artist is dependent almost entirely on private patronage. As usual, the range is from a supportive system controlled entirely by public authority and administered so as to bring art under the sway of government as sole impresario, to a system which frees art almost entirely from the shackles of government support and lets it fend for itself among private or commercial patrons.

We, in America, are ambivalent about government's involvement with art. On the one hand, art is a "good thing" and should be fostered. The opera house and performing center need subsidy ("more to the neighborhoods!"). The public square needs a fountain and the postoffice a mural; artists and writers need grants. Beauty, it may be felt, has some claim on the budget. But we shudder at the thought that the bureaucrat or committee who pays the piper should call the tune. Not only because we distrust official taste, but because we distrust, in this area at least, political motives. On the whole, we believe—although someone is always pushing for more government funding to prove we are not barbarous—that the patronage structure of the arts should be private and commercial. The artist should seek financial not political support; should work for money, for the consumer or the corporate sponsor. He should learn to make his way not within a government bureau but within a world of agents, entrepreneurs, disk-jockeys, advertising agencies, commercial networks, commercial publishers, reviewers, private wealth, and tax-exempt foundations. Thus, government control of art through direct control of its supporting structure will be minimal. This does not mean that art is without constraints. The commercial system has its own imperatives. But it does mean that we forgo government control at the point at which it is most systematically effective.

What is left is a traditional battery of control devices applied with differing degrees of vigor and thoroughness—legal concepts of blasphemy, sedition, obscenity, libel. Art is resourceful and in the running battle between artist and censor the latter seldom cuts a heroic figure. Still, censorship can be effective in societies determined to impose it, and without the constitutional commitment and protection for the freedom of communication expressed, for example, in our First Amendment.

Much can and has been said for the ultimate wisdom of freeing the arts from direct dependence on government and from systematic monitory control. But art—high and popular—is deeply consequential and, therefore, not beyond responsibility. Whether, to what degree, in what ways a polity is to support and govern art is a legitimate political or constitutional question.

The politics of awareness is now plagued by a growing and baffled concern over chemical modes of consciousness alteration. This is not a new problem, but it has new dimensions. The grape, the poppy, and their near and distant relations have left their marks upon human history and character. But history seems only a prelude to the age of massive sedation, stimulation, mood alteration, awareness or consciousness "expansion"—achieved not by taking thought ("which of you by taking thought can add one cubit unto his stature?") but by gulping, puffing, or puncturing.

> Look into the pewter pot
> To see the world as it is not.

and

> Malt does more than Milton can
> To justify God's ways to man.

seem gentle, rueful jests before the storm.

> Swallow me and you will see
> The world as it ought to be.

begins to transform the theory of illusion. On the whole, alcohol did not come wrapped in missionary metaphysical claims. It seemed simply a pleasure in moderation, a sickness in excess. But Dionysus has returned with a new bag of tricks and a whole corps of beguiling sophists to speak for the besotted.[17]

There is little point to arguing that this is all a private matter. Widespread use of mood and consciousness-altering substances is a public issue—whether the use is private or official.

It affects education. It is now a part of international politics. In most countries there is heavy, though selective control. No one, I think, can take much joy in the prospects of spiritual sumptuary legislation. But even permissive societies will find themselves, of necessity, more and more involved in the politics of this aspect of the life of the mind.

Every sustained human enterprise rests upon its appropriate condition of mind, upon a condition of awareness or consciousness. And haunting every endeavor is the fear of pollution, of infection, of the corruption of consciousness which can cripple, pervert, or destroy. There is a vast pathology of awareness, of forms of individual and social insanity, and I shall not list the ways. But among maladies, some deserve a special mention here.

Alienation and estrangement, which somehow to a sickly "sinful" consciousness seem so romantically interesting, are forms of misfortune which can support no mode of life. The furious consciousness which sees the ordinary round of social custom—and even nature itself—as an "outrage" is not a form of higher sensibility but an infantile cast of mind. The obsessed consciousness, which sinks its teeth into an ideal even as it loses its sense of proportion, parades itself as "moral" as it degenerates into ruthlessness. The disconnected consciousness—perception cut loose from its service to human action and lost in a colorful daze . . . all these are familiar ravagers on the contemporary scene.

May I reserve final comment for a condition of mind especially apparent to the privileged view from the ivory tower during the past decade? We are all familiar with the Fool's Paradise. To live in a fool's paradise is to be happy through or because of ignorance. It is a precarious happiness; built into the image is the suggestion of a fragile illusion about to be shattered by the intrusion of the truth. Transience, ignorance, euphoria—a fool's paradise.

But there is also, although it is odd that we do not speak of it, a fool's hell—a condition of misery based on ignorance, on a failure of understanding. It is a condition of awareness in which everything seems rotten, corrupt, fraudulent, empty, pointless. The world seems like hell itself, and misery the appropriate condition. A fool's hell! Why do people live in it? Out of folly. Out of failure of perception, of knowledge, of understanding. It is not the world which is at fault, but a sickened awareness.

And thus we return to the beginning, to Plato and the philosopher-king whose function it is to lead us out of the cave in which we all begin, to lead us from the perception of images to their proper interpretation and understanding, to rescue us from—whether paradise or hell—the fool's abode. Are we to say, really, that all this is not the statesman's concern?

I am uneasily aware of how much I have not touched upon as well as how much I have merely touched on. Political epistemology, if I may use that expression, is really uncharted territory. Technical philosophy in dealing with problems of perception or truth or inference has not exercised its ingenuity in exploring the political aspects of the condition of awareness. The great cognitive professions have pursued their work without too much concern for the theory of their relation to the polity and have been content with crude expressions of the conventional wisdom. Each bit of territory is well trodden. What has not been said about the great awareness institution, the mass media? What have we not heard about the cognitive professions? What can be said that has not been said a thousand times since Plato about the artist and society?

What, even, is there new to say about consciousness and drugs?

Beneath all these aspects of the production and direction of awareness lies the problem of making sense out of the normative conception of a proper or healthy awareness, for the individual and for the society. Without such a conception the institutions and the professions lose their point. It makes obvious sense to say, sometimes, that we are not aware of what we should be aware of; there are diseased conditions of awareness. In ways I have suggested, directly or indirectly, concern for the health of awareness is part of the legitimate involvement of government with the mind.

The Direction of Cognitive Energy

Every modern society must make provision for science and research; not only for a body of knowledge but for an active corps of cognitive agents—scientists, scholars, intellectual technicians. And government has a place, even a predominant place in this enterprise.

"Knowledge," as everyone knows, "is power," and the pursuit of power is always of interest to politics. If we forget this, as we are inclined to do, it is because the motives of the prime movers may seem remote from ordinary political (and even utilitarian) motives. This is especially the case since the scientist has, in his own mind, if not altogether in popular judgment, separated himself from the magician, whose quest for power is more nakedly asserted.

The reputable governing motive is taken to be curiosity—it all begins in wonder, or, in a more sophisticated version, aesthetic thrill—the thrill of finding the ratio, of seeing the unity behind the appearance of multiplicity, of grasping the order behind the face of chaos. Science, we are told, is beautiful. That it may also be useful is, to the purist, a contingent side-effect, and to be overly concerned with *that* is to fall away from the spirit of the elect. Others may exploit the utility of scientific insight, but the true scientist himself is driven only by his peculiar thirsts. A familiar drama. Something followed by its votaries for its own sake, needed and sought by others for their own mundane purposes; the freedom of science, and social necessity.

The freedom of the scientist to follow his own bent has, as might be expected, found support in the familiar free-enterprise model. Each individual seeks to satisfy his own curiosity and in doing so contributes to the maximization of the gross knowledge product. Arrowsmith, disgusted with research bureaucracy, returns purified to the little laboratory in his garage. The grip of this model is still strong and we fall back on it almost instinctively when a scientist's autonomy is threatened, although its appeal is tinged with nostalgia. It is, I think, the wrong model. For one thing, Arrowsmith cannot buy his equipment and can hardly afford the journals. The world in which much of science works is not the macro-world, the world, in Stephen Pepper's phrase, of middle-sized fact. It is a world into which one can penetrate only with massive and complex instruments. The tools must be provided and the provision justified in terms other than providing thrills for scientists. Something may remain of the spirit of small-business free-enterprise in the life of the mind, but "the marketplace of curiosity" no longer reveals the basic plot. If we want an explanatory first-approximation model I suggest that we turn, instead, to the idea of the city and consider the research-oriented university—the typical center of modern science—as a new kind of city.

In a fanciful history, civic life grows around a focal point of power—a fortress on a hill, a market at a crossroad or port, a ruler's court, a seat of religious authority, the factory, the bank or financial district. As a new form of power develops,

it establishes its central citadel and adds its characteristic sub-
urbs. The power develops its priests, acolytes, and clerks, and
its city grows by accretion of its servants, its clients, and its
camp-followers. It exacts levy from its hinterlands or ex-
changes benefits for support. It is magnetic in its effect, the
power-seeker's El Dorado. And when its power is exhausted
or overshadowed, it withers into a ghost town, an animated
museum, a tourist stop.

It is now the University's turn. It is the seat of knowledge—
for modern civilization the indispensable form of power.[18]
From small beginnings, the handful of scholars in a humble
building in a field has grown into a powerful company in the
midst of a booming campus dominating a once-sleepy college
town. This central city draws its sustenance from the outside
world—support sometimes gratefully bestowed in an awareness of benefits received (and to come), sometimes reluctantly,
as if tribute to warlords or to bandits who hold children as
hostages. From this new Camelot sallies forth a stream of missionary consultants, fresh from the round-table, returning
with tales of untamed lands, harrowing accounts of ignorance and disorder, of snarls unsnarled, with news of un-
spoiled lakes and inns, with modest displays of loot.

Internally the city has the familiar features—its structure of
chancellors, deans, chairmen, senates, assemblies, committees,
departments, hierarchies, tenure-holders, candidates, degrees,
struggles for precedence and priority. And the familiar types—
apprentice and master, monastic and worldly, quietist and activist, traditionalist and radical.

It is within such a city that the life of science and research
takes shape. Research does not just happen to take place
there; it is its way of life (as fishing is the way of life of a fishing village). Curiosity and the pursuit of cognitive satisfaction are here habitual and institutionalized, professional, not
impulsive and amateur, systematic, and accommodated to a
structure determined, in good part, by other considerations.

The cognitive energy which flourishes here does not provide its own direction. It is neither, altogether, self-governed nor ungoverned, although it may sometimes seem to be both. There is a politics of research, and it is an essential feature of the enterprise—a fact obscured by the tendency of scholar or scientist to be dominated by his own work and to impatiently scorn what makes it possible as mere administration. Let us consider, therefore, some of the general features of the politics of cognition, the constitutional structure, the rules, the policies of the research city.

If there is to be a community of men and women able to devote itself to a life of research, there must be support which not only provides facilities—libraries, laboratories, equipment, materials, services—but which is sensitive to problems of motivation and temperament and provides for an attractive way of life.[19] This support is negotiated in terms of utility, long or short range, from both private and governmental sources. Spectacular achievement in time of war or emergency, visible contributions to technology—agricultural or industrial—the training of classes of professionals, strengthen the claims to support.[20]

The problem—the danger—is that support may not be sufficiently free from strings, from specific direction. Society may want *this* problem solved now; then *that* one. Each donor has his own urgencies and priorities. But the guardians of the research city understand better the necessities of a continuing and fruitful life of research and, while not unwilling to be useful about what others regard as urgent problems, attempt, as far as possible, to determine and protect their own agenda. This is not merely out of a self-indulgent desire to pursue one's own interests, but out of the conviction that significant research or science is not always the result of ad hoc responses to other people's questions.

The basic issue, then, is over the terms of support, over the degree to which it permits independence from direction by the donor, public or private. At one extreme would be a system in which every grant of support was tied to a donor-determined problem; at the other extreme would be a system in which the institution received only general grants which it could use entirely as it saw fit. Within these extremes there is a constant struggle between pressures toward subservience and the drive for autonomy. This struggle, together with the question of the level of support, constitutes the core of the external or foreign relations aspect of the governing of the research city.

The university (as the "research community") has developed an armory of devices. It may develop a distinction between institutional and project support and negotiate a hard, minimal structure of institutional support. It may siphon off a portion of every particular grant or contract for overhead or institutional purposes. It develops skill in formulating proposals in terms which attract support and in subtly transforming by interpretation the purpose of a grant into something closer to its heart. And it infiltrates the donor-advising system. Research policy is highly technical and government and non-governmental donors need expert advice. They draw heavily upon the advice of scholars and scientists. Thus, the perspective of the research community is represented at the source of external support and helps to shape or moderate policy. The aim of all this is to spend other people's money as you think it should be spent, without endangering its supply.[21]

Perhaps the chief device by which a society may ensure a fruitful autonomy for its institutions of research is the decentralization of support. At present, the university in America enjoys support from both state and federal governments and its many relatively independent agencies. It is the beneficiary, as well, of tax policies which permit vast accumulations of funds in the hands of foundations and which encourage private support. The proliferation of sources of support, actuated by different policies, increases the independence of the research institution, frees it from a precarious dependence on a single source.

I do not really mean to suggest that university and society are fundamentally at cross-purpose. They are not. But misunderstanding, impatience, and resentment are inevitable, and the political challenge is to work out the dependence-autonomy relation so as to benefit everyone. For behind this problem lies the real question. How should research be done? What problems and in what order? How basic? How applied? Who is to determine the actual deployment of research energy? Or rather, how is it to be determined? The research community has its habits, its professional principles and inclinations; the supportive community has its needs and expectations. And each has a favorite myth. The academic myth is that there is a proper order of research—that left to its own devices the community will pursue the truth as required by some sort of natural schedule, conquering knowledge step by necessary step—in short, that the pursuit of truth has its own inherent order. An order which should be interfered with as little as possible by pests with practical problems. The counter-myth is that cognition is always in the service of action and that until the mind is given a problem by the exigencies of "real life" it is essentially idle. Keep it busy with real problems and it will organize itself to deal with them; leave it to its own devices and it becomes impractically academic. In short, it *needs* externally imposed tasks and should respond to them as they arrive or are imposed. That is, the proper order for the life of research is a priority list of practical problems—politically ordered.

This may be a slightly caricatured version of the abstract politics of the True and the Good. The quarrel between the autonomous and the instrumental conceptions of knowledge is an ancient one. It is hardly likely to be settled neatly or dogmatically by a research community, truth-loving though it be, supported by a polity in search of the good life.

In fact, when one turns from the "foreign policy" of the research city to consider how it runs its internal affairs, how it determines how its discretionary resources are to be deployed, one finds politics of a distinctly lower form than is evoked by the necessity of winning external support. That is to say, it is a politics of allocation among vested interests, governed by the principle of live and let live, by a reluctance, elevated to the status of principle, to judge the significance of work in another's field or collectively to question the value of a part. The favored policy is that everyone should have more, or at least as much. If that is not possible in one of those unfortunate crises of support, peer-group, pork-barrel, democratic government may be overshadowed for a time by an autocratic administration called to make some hard, rude, insensitive decisions in the crudest possible terms—i.e., this is more important than that. And why? How dare one say that? Support. Present support and probable future support. The country will need more of this . . . and will pay for it. This is not only a perfectly good reason, but in the absence of anything but a vested-interest plea, almost the only serious reason in the field.

If, in short, one looks for something higher than utilitarian politics in the government of the cognitive energies of the research city, he will not find it. There is no great spiritual master-plan. Only another city with its traditions, institutions, habits, and politics. Nevertheless, it *is* the cognitive city; it is the place of the mind; it is a seat of power and it can, if we are fortunate, negotiate the terms which allow it to work in its own way—incomprehensible as that may seem. Incomprehensible as, for example, is "tenure."

Tenure is a crucial academic institution which is easily misunderstood. It is not simply "security of employment." A tenure system is, in fact, about the only system in which a public employee may be terminated for less than "cause." It normally involves more than a half-dozen years of pre-tenure service which may end without tenure simply because the institution is unwilling to make a tenure commitment in that case, even though performance has been diligent and even satisfactory.

Once granted tenure, however, the scholar is considered to have served his apprenticeship and the question of his basic competence to do research in his own area under his own direction is taken as settled and not to be routinely reopened. To hold a tenured position is to have the strongest claim to continuing institutional support. One is not to be fired in order to be replaced by someone "better." The tenure holder may seek, but does not depend upon, special external support. He can go his own way doing unsupportable things. He can go out of fashion and provide the institution with some cultural lag. He cannot be ordered to redirect his attention or energy. He can decline to share the enthusiasms of his colleagues and can amble after the truth as he sees it, heedless of anyone's longing to fill his slot with a hot replacement. By virtue of tenure the university has a built-in resistance to over-responsiveness to both external support or donor pressure and to internal collegial fashions. To some extent, research can go its own way, shielded from immediate judgment.

The individual who has tenure does have security; but the institution has something even more important—freedom from an overwhelming burden of destructive academic politics. This, I believe, is a point which dawns rather slowly. It is normally supposed that the point of tenure is to protect the researcher against external or administrative sanctions. It does that, of course. But the real threat to scholarly independence is not the external public raging against the lone dissenter, nor even the vested interest dropping a sinister money-laden hint, but collegial pressure. Suppose that, without

tenure, a scholar's reappointment depended upon the recommendation of his colleagues. University life would become
unbearable. If that seems a bit cryptic, it is perfectly clear to
any tenure member of a research faculty. The real point of
tenure, from the governmental point of view is that, at its
own price, it removes a whole range of decisions from the
agenda, protects the institution from some "reasonable" demands, diminishes its responsiveness to plausible, short-range
urgency, and, in general, increases the ungovernability of the
research community. Tenure is a great invention and will, if
we are fortunate, survive the occasional proposal by the naïve
reformer that it be abolished.

To understand the politics of cognition it is necessary, I
have argued, to understand the way of life of the academic
city, the way in which it decides what to do, the way in which
it is moved both by external forces and by internal habits.
When this is understood, the question, if it can really be
called that, of the legitimacy of government's concern with
this province of the domain of mind will seem strangely unreal.

It is, nevertheless, characteristic of the academic profession
that in its conception of itself these political factors virtually
disappear from the picture. The heart of the drama is, for it,
the seeker after truth and his quest; it may care for little else.
But the actor does not furnish the stage or even write the
play. A general diminution of public support forces the institution into priority decisions. War and cold war will result
in massive funding for particular areas of science; a shift in
the situation and other areas will flourish. Now the atom,
now space, now urban problems, now ecology. The effect on
the university is obvious. Some departments expand as they
can now afford facilities and personnel. Fellowships and schol-
arships become available. Students find a boom going; pres-
tige, careers, opportunities now lie here rather than there.

For most students it is not a case of giving up what one really wants to do because it is only possible to do something else. It is, rather, a case of focusing a vague or general interest in the particular way encouraged by the situation. He adjusts his path to the availability of opportunity, and lo!—society has, for a generation at least, a fund of intellectual energy committed to a particular area. And, on the whole, the scholar or scientist is conscious only of the fact that he is pursuing his own interests and is being supported in doing so, with little sense that he is being "governed." But, offstage, the political decision that we need more of this and less of that shapes the pattern of opportunity and, in due course, the shape of knowledge itself.

This discussion has focused on the modern public research-oriented university as the central city of knowledge. Not all research goes on there, but the same analysis is generally applicable. There are great private universities, but their stories differ only in detail. There are also research centers which are not university-located—the industrial or commercial or governmental research laboratory and private institutes or study centers. Typically, its members are university-trained, university-connected, or, perhaps, university refugees. They are almost university outposts. They also have support problems and the character of research there may be even more responsive to donor direction. These institutions may be free of some of the constraints of the university but they have constraints of their own. The inescapable fact is that science and research are, today, institutional in character. And even the most utopian intellectual colony sadly rediscovers the politics of the mind.

From this general consideration of the cognitive enterprise seen as a social institution involving government, I turn now to some questions of social policy as they impinge upon the pursuit of knowledge.

The pursuit of knowledge may be the pursuit of a "good thing" but it does not follow, on that account, that it can claim a sovereign autonomy transcending any limitation by considerations of social utility, or morality, or religion. The pursuits of other good things—nourishment, pleasure, etc.— are subject to constraints and there is no obvious reason why the pursuit of knowledge should be regarded as a thing apart. It is important; but its pursuit is haunted by fallibility and error, and its fruits, in rash or evil hands, may destroy us all. We may, I think, dismiss out of hand the view that since knowledge is good it is always sinful to interfere with or limit or constrain its pursuer—to bring him under government. Even science is subject to the law.

The clash between basic social assumptions and science is an old story. The schoolboy examples are the attempts to protect geocentric or homocentric or theocentric conceptions, believed by their defenders to be the necessary foundations of "civilized life as one knows it," against subversion by the Galileos and Darwins of the world. In the old historical cases it may be possible to tell the good guys from the bad guys without the aid of a program. Contemporary cases are more puzzling. Should science accept the necessity of operating within the framework of national socialism? Or the principles of Marxism-Leninism? Or of Chairman Mao? Or, even, of democracy? Must genetics obey the equal-protection clause? ("But we're born smarter," he muttered as he was hustled from the hall.)

It is obvious, if we linger a moment over the storm about research in heredity and intelligence, that the issue is not entirely over the quality of research. There is argument over whether certain claims are false and hypotheses ill-formulated and unconfirmed. But the real heat is over the threat posed to a fundamental dogma, to "all men are created equal" read in fundamentalist terms. If science may show that men, that ethnic groups are not . . . No, stop, we don't want to know, we don't want it known, it just can't be, let's stop testing, let's not provide ammunition for scoundrels, let us de-tenure, un-fund, boycott, unpublish, and protect our efforts to create a better life for all. I dwell on this example out of some malice toward those who seem so unreflectively and unsympathet-ically sure that they know the right side in the perennial con-flict between enlightenment and dogma.

The fear of knowledge is not entirely a barbarian or a pop-ulist vice. Knowledge is power and whether, like other forms of power, it tends to corrupt or not, we may be fairly corrupt to begin with. Or foolish. There is no built-in fail-safe guar-antee against misuse. Scientists themselves, contemplating the use to which their work may be put and feeling "responsible" (not that they can claim, let alone display, more than a nor-mal share of political virtue and insight) may be attracted to the possibility of self-imposed restraints or concerted refusals to develop the kinds of knowledge which, in the state of the world's political life, they deem dangerous. Can we stand more explosive power? Who can be trusted with the tools of genetic engineering, or with those big computers?—apes in a power house with all those buttons! Can we agree, by treaty, to abstain from research on germ warfare or nerve gas? Is there no way to stop Pandora?

We have known for a long time that knowledge is not the same as wisdom and that the former may outpace the latter. Is it legitimate to try for forms of moratoria, to slow down the pace of research and discovery? Or are we forever in thrall to the self-interested slogan that there are no problems created by knowledge that more knowledge will not cure?

Cognitive energy, like any form of consequential energy, is, in principle, subject to social control. Its level and direction are, as I have argued, affected by institutional politics. But beyond that there are devices by which a society may control it more pointedly. I mention but do not pause over regimes which are in a position and are quite willing to simply command that something not be done. No research along Mendelian lines! No Jewish physics! Or which draw the line at publication or promulgation of heretical doctrines by systems of censorship and limited circulation. A country like the United States has largely stripped itself of the power to prohibit, directly, prying into a particular field of knowledge. It encourages or discourages chiefly through the provision of support, and even that control is rather tenuous and falls far short of the ability to impose an orthodoxy upon its scientific community. On the whole, the conception of formally forbidden areas of inquiry is repugnant to us.[22]

Procedural limitations, on the other hand, are quite familiar and are likely to increase in importance. There have been persistent attempts to protect animals against pain and destruction in the course of experiments, with limited legislative success. The use of human beings as experimental animals is not unknown in history and frequently evokes revulsion and prohibition. But the more subtle problems fall a bit short of the mad scientist and his hapless victims. Considerations of dignity and privacy appear increasingly to demand the imposition of limits upon scientists (especially social scientists) who are not inclined to impose limits on themselves.

One tendency is to extend the notion of "informed consent" from the medical context in which it was developed to broader areas of experiment and study in the psychological and social science areas. The difficulty is with research which depends upon deception or upon the unawareness of the person who is being studied. It may burden some research to insist on meaningfully informed consent. On the other hand, not to require such consent subjects persons to study in ways they might regard as invading their privacy and even as potentially damaging to their status or self-respect. The use of data, obtained under conditions of confidentiality, for purposes of study in other contexts is being questioned. We may, in short, have to face new problems about the rights of the involuntarily studied. We may even have to deny ourselves some knowledge that might be useful for business, or political, or educational, or law-enforcement purposes. Principles of voluntariness, of informed consent, of privacy must be reckoned with.

It is, I think, rather fitting that the hard drive to inquire, the lust to know, should find itself checked and baffled by such soft notions as consent, dignity, and privacy. It has been on a long and glorious rampage. It has dispelled much ignorance and illusion, flouted orthodoxies, trampled on the sacred, defied taboo, freed itself, almost, from check by any external authority. What can withstand it? The lone, stubborn individual who doesn't care to be looked into, a sense of offended dignity, a tattered principle of privacy growing ever larger and more powerful in an impersonal world—privacy, the looming antithesis to the public and the social?

I have tried, in this section, to show that there is, and must be, legitimate political concern for the condition of the mind whether we consider the problems of current awareness or the problems of the direction of our cognitive energies in the longer-range pursuit of knowledge. In both cases we need to be aware of the basic institutional structure supporting the condition and the activities of the mind—the mass media, the arts and the cognitive professions, the institutions of research. It is clear that the community, acting in different ways through its political institutions, has a hand in determining the structure of institutions so crucial to the life of the mind. But beyond all this we must be aware of the specific policy questions and the limitations, in the name of policy, that are imposed or that we may seek to impose on the operation of the institutions that affect us so fundamentally. The discussion may take on more concreteness as I now turn to consider the two great institutions of the mind—the school and the forum.

3

Government and the Teaching Power

The art of government, in as far as it concerns the direction of actions of persons in a non-adult state, may be termed the art of education.   JEREMY BENTHAM

Consider for example the case of education. Is it not almost a self-evident axiom, that the state should require and compel the education, up to a certain standard, of every human being who is born its citizen? Yet who is there that is not afraid to recognize and assert this truth?   J. S. MILL

The social "present" is generationally thick. Although we may think of child, parent, and grandparent as representing the future, the present, the past, they are all here now; and while each is at a particular stage and may, in his self-centeredness reinforced by peer-group consciousness, think of himself as the moving center, to the dispassionate observer peer groups come in stacks, and the here and now is many-layered. Lost or isolated generations hold a special interest—as when the Pied Piper removes a layer, or when a retirement village cuts itself off, or when a horde of children rediscover Beelzebub on an island or campus—but, normally, the social present is thick with generations.

This circumstance—that the individual lives through stages and is always at a particular place, while the society at any particular time has all stages present—has a profound bearing upon political theory and especially that part of political theory which concerns the relation of government to the mind. It means, I believe, that no single set of principles is adequate to the governing of the entire range of stages and that the basis and scope of authority within a single society may vary drastically over its different stages. Let me elaborate a bit.

There is a strange passage in the *Republic* in which Plato with deceptive diffidence offers us a myth which, he says, everyone should, if possible, be brought to believe. It is, apparently, a ridiculous tale and it is greeted as such and may seem so to us, until we realize that it is among the deepest accounts of our creation. I quote a relevant part. Everyone is to be convinced, when he reaches adulthood, "that all our training and education of them, all those things which they thought they experienced were only dreams. In reality, all that time they were under the earth, being fashioned and trained, and they themselves, their arms and all their possessions were being manufactured, and when they had been made quite ready, this earth, their mother, sent them up to the surface. Now, therefore, they must watch over the land in which they dwell, as their mother and nurse, and defend her against all invaders, and look upon the other citizens as their brothers and children of the same soil . . ."

The proposal is, of course, that we be brought to understand ourselves as children of the polis, as the political animals that we are; that we understand our nonage or childhood as part of the process of the birth of a person, a fetal stage during which we receive our equipment—the language, habits, culture—without which we cannot emerge from underground, from limbo, from the social womb, from the childhood that, when we recall it later, seems like a strange prenatal dream. We are born in stages and, for the crucial stage, the polis is parental. We are, accordingly, siblings, brothers and sisters, children of the polis, polis-animated. The ridiculous tale is utterly true.

But generation is painful and the generations may forget the traumas of birth. In a familiar crisis of identity a generation may deny its generator and deny that it was generated at all. "Nonsense!" says Satan when the faithful Abdiel reminds him of his creator; "As long as I can remember I was there." Surely, no one who endured life in the American incubator of the sixties can forget the sophomoric cry of generational revolt: "We are a self-created angelic generation, immaculate, untainted by parental sin, and we will remake the world [Ah, Pandemonium!] in our own image." Nature, when she is bored, imitates art.

The point of all this is so obvious that we need constantly to be reminded of it. A society is not an undifferentiated heap of individuals, equal, at the same level of "authority" and "right." It is a continuing entity, continuously regenerating itself, always pregnant, always with a generation in limbo, always with a part of itself in a condition of tutelage. The distinction between minor and adult—however much we may be baffled by borderline problems, by demands for adequate criteria, by administrative difficulties—is fundamental and inescapable. There is no society which does not recognize the distinction or mark, by some rite of passage, the movement from one condition to the other—the achievement, as we would say, of the age of consent. No single set of principles can adequately govern both minor and adult; we need both caterpillar principles and butterfly principles. *Republic* is a discussion of the raising of children; *On Liberty* is a discussion of the governing of adults. They are complementary works about different generations. John Stuart Mill would have been horrified by the application of the principles of *On Liberty* to children.

The special authority of a community over its emerging generation grows naturally out of the minor-adult aspect of the human situation. The conception of the community as the womb of the person puts the matter beyond the question of merely formal political authority and into the domain of parental function. The nurturing of children is to be seen, on the one hand, as the developing of persons and, on the other hand, as the process of social self-preservation and renewal—the re-embodiment, the reincarnation of the parental culture through the creation of the individual in its own image.

The natural right of self-preservation lies behind not only the traditionally asserted powers of war or defense, but also the universally claimed right of the community to shape its children. More fundamental and inalienable than even the war power stands the tutelary power of the state, or, as I shall call it, the teaching power.

The *teaching power* is the inherent constitutional authority of the state to establish and direct the teaching activity and institutions needed to ensure its continuity and further its legitimate general and special purposes. It is rather strange that a governmental power so visible in its operation and so pervasive in effect should lack a familiar name. The Supreme Court refers to the power of the state "to prescribe regulations to promote peace, morals, education, and good order of the people . . ."—a power, it adds casually, "sometimes termed its 'police power' . . ." But it will prove useful if we separate out the school and call the power of government which comes to focus there by its own appropriate name. The teaching power is a peer to the legislative, the executive, and the judicial powers, if it is not, indeed, the first among them.

In a federal system we may have questions about the location of the teaching power. It is generally assumed, in the United States system, that powers not delegated to the Federal Government are reserved or retained by the states and, accordingly, that education is primarily a state matter, although subject, as are other state matters, to federal constitutional constraints and enjoying, under a variety of pretexts, federal support.

The teaching power is not limited in its scope to children or minors, although I stress that aspect. To exercise the teaching power is to make claims upon attention and to subject mental and physical energies to discipline. This is done in various ways. We may, as with minors, enforce attendance in accredited schools with required curricular elements for a number of years. But beyond compulsory schooling we may provide for general, professional, and vocational education. And here the idea of compulsion gives way to competitive claims to opportunities as we scramble for places in schools of medicine, law, engineering, etc., provided, in part at least, by the state. Beyond this voluntary relation to the teaching power by adults there may even be compulsory or quasi-compulsory relations in corrective, penal, or therapeutic situations. Thus the teaching power ranges over minor and adult, in voluntary and involuntary ways, for purposes ranging from the sheer necessities of survival and continuity to the enhancement of the quality of social services and individual lives.[23]

The state's claim to the teaching power may be asserted in a strong or a weak form. The strong claim is that *the* teaching power is vested fully in the state and is to be exercised exclusively by agencies of government or, in a variation, by licensed, authorized, supervised non-governmental institutions which operate within governmentally determined policy. In the strong view, non-governmental institutions—religious, commercial, private—in the domain of education exist not by inherent right but on tolerance or out of considerations of policy.

In its weak form the assertion is that the state is one of the legitimate claimants to *a* teaching power; it does not enjoy monopoly or even priority; but it may enter the field. This view would support an extensive proliferation of public education institutions, but governmental exercise of the teaching power would have to accommodate itself to the equal legitimacy and even independence of other teaching institutions. The weak assertion might carry us to the requirement that all be educated to a certain level; but not that we necessarily struggle in government classrooms.

The strong and weak versions pose, in this context, the bitter controversy between unitary sovereignty and pluralistic theories of the state.[24] Fortunately, it is not necessary to resolve this dispute at a theoretical level in order to explicate or develop the notion of the teaching power. Nor need I, nor do I, assume the burden of defending the assertion of the teaching power in its strong form as appropriate for contemporary America. (But see note 24.) We are deeply involved in the problems and politics of education and, I believe, the missing or neglected conception of the teaching power will clarify and aid our understanding. But it may well be the weak version which is implied in our theory and practice. In any case, that is a version more likely to be hospitably received; is compatible, if it should prove necessary in the end, with the strong form; and is sufficient to establish government's legitimate involvement with the mind. There is, even on a grudging view of our constitutional system, *a* teaching power, exercised by government, sustaining a public school system, whose existence or legitimacy is virtually beyond challenge.[25] That is not to say, of course, that the teaching power is boundless in scope or untrammeled in its exercise. Government is permitted, in its exercise, much that goes beyond what it is permitted in the governing, as we shall see, of the forum. But it is, like any governmental power, subject to the general and particular constraints which are the conditions of constitutional legitimacy.

The teaching power is vested in a structure of offices and institutions. The Constitution of the State of California, for example, states (Article IX) that "A general diffusion of knowledge and intelligence being essential to the preservation of the rights and liberties of the people, the Legislature shall encourage by all suitable means the promotion of intellectual, scientific, moral, and agricultural improvement." (I love that statement for its sheer sanity, its uncomplicated directness, its casual profundity. It deserves a place, in our reflections about government, beside the First Amendment.) The California Constitution proceeds to provide for educational officials and institutions, and what evolves, here as elsewhere, is a formidable array of state or public schools—primary through university—ranks of administrative officials—principals, superintendents, chairmen, deans, provosts, chancellors, presidents, boards, councils, regents, trustees—and masses of teachers or "officers of instruction."

There are charts for everyone's taste, and the normal habits, tensions, and problems of bureaucratic life—hierarchy and autonomy, centralization and delegation, expansion and retrenchment, tradition and innovation—all bearing, although sometimes remotely, on who teaches what and to whom. But the politics of the teaching power is not confined to its internal organizational dimension. The past decades have brought into unusual prominence the surprising range of problems which haunt the exercise of the teaching power. Taxpayers, politicians, and courts struggle over the level and distribution of support; parental, ethnic, and neighborhood organizations press claims for equality and dignity; students present demands; teachers struggle to respond to new expectations and claims and to preserve their professional integrity; theorists and ideologues rush in with advice. Everyone organizes to parley or to fight as the society, uncertain about itself and its future, places heavier burdens on the school. It is not that the school has suddenly and improperly become "political," but rather that the natural, inevitable, and legitimate politics of the school has become more urgent and more visible. The political struggle over education in all its aspects is, without exaggeration, among the most significant of our time. It is a bitter conflict over unavoidable issues and the stakes are high.

It is within this prosaic but turbulent framework that it becomes possible to understand the otherwise exotic conception of academic freedom. Immersed as it is in a sea of pressure the teaching branch of government claims the power and discretion it must have if it is to do its work. The legislature has its privilege, the executive has its prerogative, and judiciary has its independence. Each branch claims, within a system of due process, the freedom, in its own domain, necessary for the integrity of its function. Academic freedom is simply the extension of the principles of separation of powers and due process to the teaching power. A violation of academic freedom is a breaching of the constitutional structure of the academic branch of government.

If it is not that, it is difficult to make sense out of it at all. The academy is not a subsidized enclave within which teacher or student may do as he pleases. Teachers are not ambassadors from another country enjoying extraterritorial privileges; they are not licensed to steal children. But the absence of a working conception of the teaching power encourages misconception and makes academic freedom difficult to understand and explain. Teachers are, after all, hired, and sometimes fired; texts are selected and rejected; courses are approved or discontinued; curricula and requirements are established and changed; teaching methods are authorized or disallowed; students are admitted or turned away. When controversy develops at any of these points and flares into an academic freedom case, what is the case? Surely not that such things are done at all, but that something has been done by the wrong person or tribunal, or by a flawed process, or in violation of the relevant criteria and rules. The teaching power as a branch of government has, and is part of, a constitutional structure. Its integrity, its place as a power in a system of powers subject to the separation of powers, is defended under the banner of "academic freedom," which claims for it what "judicial independence" claims for the judicial branch.

The scope of the teaching power is so formidable that the problem may appear, in the end, to be less how to protect it than how to limit it. It is not always diffident in asserting itself. Consider a classroom in a public primary school in a typical American community: a captive audience, involuntary reading, involuntary writing, involuntary reciting, involuntary revelation of guilty ignorance, all backed by the power to classify, grade, promote, fail and expel—sanctions which, in terms of consequences for one's life, make pale indeed the transient chidings of the judicial power.[26]

The teaching power has its share of the general problem of government; it is another institutional setting for the study of politics and public administration. But it has as well the peculiar problems native to its special character. Its function is teaching; its unique functionary is the teacher.

Teaching must be seen both as an art and as an office. That teaching is an art is generally granted, although it works so mysteriously and assumes so many guises that we often attribute its fruits to luck or simply to not interfering with the inherent powers of the learner (forgetting that teaching is partly a strategy of non-interference). Teaching activity is so pervasive that much, perhaps most, of it escapes self-consciousness or identification with the teaching role. But it rises, at some points, to awareness of itself, finds its heroes and masters, cultivates its lore, and achieves the status of art and profession. As a profession it is properly seen as a fellowship entrusted with guardianship over a social function. And, in due course, as the function is provided for by the politically organized community, the profession finds itself enjoying and chafing at public office.

In this sweep from activity through art and profession to public office there is tension and confusion at every point. Much teaching, as is much that is called cooking, is overdignified in being called an art, and seems in the one case to spoil learning as in the other to spoil food. Much of the art is, in novel and creative modes, denied the imprimatur of the guild or profession. Some of the profession escape or seek escape from the constraints of the formal teaching office. And some of the holders of the teaching office, notably university professors, deny flatly that they are really officeholders or public agents at all. The art is restless in office and resists the curb. Even Socrates is enigmatic. He was, of course, a master of the teaching art to which, he believed, he was called; but he thought of his teaching as the exercise of an office in the service of the polis. His suggestion, when challenged, was that he be appointed and even paid. Athens declined the opportunity. An unwise decision, Socrates thought, but one whose authority he acknowledged. The nuances of that episode still puzzle us and the message can be misread. But the teaching power, at any rate, brings the art into office.

The office is a sensitive one and involves, as do medicine and law, close and confidential relations. There is dependence and there are frightening possibilities of misdirection, exploitation, and betrayal. The teacher is an agent in a position of trust, and it follows naturally that access to the teaching office is quite properly restricted. Merely "wanting to teach" is not enough and is often, in my experience, a surprising sign of unfitness.[27] One must be admitted to a profession which maintains itself by co-option. There are systems of candidacy and apprenticeship. Fitness must be established. There are not only technical qualifications but a broader range of considerations having to do with the ethos of the role. The latter are very important, very "obvious," and yet very difficult to translate into administrative criteria. So difficult, in fact, that the basic principle tends, too easily, to get discredited. The argument is that technical competence in "the subject" is all that is required. Because, presumably, the teacher will just teach *that* (mathematics, geography) and will not bring his private ideology or philosophy into the classroom. When, however, the obstreperous teacher, insisting on "wholeness," "integrity," and "conscience" does bring all that in, his "right" to do so is defended (unless, of course, he is a racist) in the name of free speech and the marketplace of ideas. The utter inappropriateness of these notions applied to a captive audience of minors in a school is so obvious to any sane person that the existence of this syndrome is believable only because it can be observed. In response, the simple-minded (but sane) fall back on notions of loyalty and orthodoxy as the appropriate spiritual complement to the teacher's technical competence, and in times of crisis we suffer populist demands for loyalty oaths and the purging of subversive teachers. This program is not pursued with the zeal and thoroughness displayed by "revolutionary" regimes, which do not fool around with idiosyncratic teachers, but some martyrs may be created. And, among the academic freedom issues, the question of the autonomy of the profession may be tentatively raised.

Whether the profession governs and polices itself or shares with laymen the authority to judge qualifications, to appoint, to discipline, and to exclude is a matter of some importance. The profession by instinct is against lay intrusion into the heart of its affairs. It is, it says, the best judge of fitness and it can, it claims, best handle its own disciplinary problems. The first is generally conceded; the second is treated as a joke or a scandal. Whether we consider the legal, medical, law-enforcement, or teaching professions, the professional capacity to tolerate marginal freebooters often seems excessive. But while autonomy claims receive some deference even in the case of teaching, where a profession works largely within a public institutional setting lay influence looms larger. The teacher may be a member of a profession; but he is also, in the usual case, working in a school—a public agent subject to a measure of political control. The teacher, as a wielder of the teaching power is ultimately answerable to the polity.

Life within the orbit of the teaching power shares the quandaries of life within any great bureau—the struggle to preserve the central vision against the corrosive effects of institutional inertia, habit, sloth, ambition, and time-serving cynicism; to preserve integrity against the pressures of institutional necessity, external and internal politics, colleagues, clients, and critics—whether seen from the point of view of the lower-echelon maverick or the senior establishment guardian. It has, additionally, the problems of a profession asserting, in its pride, the claim to autonomy while working within and subject to the constraints of public office. All, we must now consider, to what end?

The teaching power is primarily responsible for those institutional processes through which individuals are developed, recruited, and prepared for social functions or, more broadly, for life in a particular society. To the already familiar Platonic images of womb and cave let us add that of the great ladder, accessible to all, which each person climbs to the height of his powers, to the social and functional level suitably his. The teacher governs the ladder which, in its full Platonic or Jeffersonian reach, involves universal schooling and careers open to talent, with mobility, regardless of parental status, determined only by ability and character. Societies fall short of this ideal in characteristic ways, but it is surprising how rarely present-day societies repudiate the ideal itself—democracy and dictatorship alike. All seek continuity through the developing, husbanding, and directing of the energies of the mind. All, that is to say, must engage in education.

This is not a treatise on education but rather a squint at it from a particular perspective; not that of the individual learner but that of the teaching power itself, considering its task. That task, in its most general terms can be seen as *development in a context of initiation*. It is the combination of "development" and "initiation" which is crucial and it is the failure to temper the one by the other which breeds both individual and social monstrosities.

Development is a familiar educational idea and its very use is a protection against the errors of cruder notions—the potter's shaping of clay, the filling of bottles, the stuffing with input, the conditioning of responses. Its attendant notions are more organic—cultivating, nourishing, unfolding, growing, strengthening, ripening—and, as any teacher recognizes, fundamentally appropriate. But development is only part of the story and, on its own, may generate anarchic or individualistic aberrations—the worship of the purely inner light, eccentricity, self-centeredness, solipsism. Development, yes; but in a context of initiation. For education is also the initiation into the ongoing activities of a culture, its arts and enterprises, its fellowships and pursuits. The great and universally applicable example is language: the development of one's linguistic powers is an ever-deepening initiation into a particular set of cultural habits. And what is so obviously true of language is true of every human art, activity, power.

The teaching power has, so to speak, a double focus. One eye is on the particular student—his special bent, his character, his talents, his potentialities, and even, for what it signals, his likes, dislikes, desires. The other eye is on the needs, the tasks, the opportunities, the practices to which the student must, in his development, be led, to which his energies must be yoked. Teaching is not only developing; it is recruiting and initiating as well. The teaching power's task is not so much to transmit culture as to continue it.

Thus, the teaching power, deployed at a crucial front, deals routinely with the generational crisis. Or rather, normal generational tension becomes a crisis when, for one reason or another, the teaching power is unable to take the inevitable challenge in its stride. Initiation into an ongoing enterprise involves, to some degree, a confrontation or encounter with the given. Recalcitrance, rejection, rebellion are, as we know, normal aspects of the complex response. The desired outcome is a well-tempered involvement, even commitment; failure, for the teaching power, for the society, and, most disastrously, for the individual, appears as alienation or estrangement—the deepest, although sometimes fashionable, of social diseases.

While it is obvious that the state acting through its teaching power is necessarily and legitimately involved with the mind, special features of that involvement are not always appreciated. The school is not a public forum and it is not governed by the same principles; children are not adults and are not governed by the same principles. If we grasp this we can begin to understand the distinctive exercise of the teaching power and not gape in foolish horror at the discovery that the school is neither a town hall nor an intellectual fair. It has its own version of due process. But it has unusual power to create and protect a special intellectual environment within which it may determine the mind's agenda and cultivate its proper manners.

To begin to understand the school and the teaching power, therefore, requires that we begin by taking two simple steps;

the first takes us over the cliché that marks off the realm of intellect and spirit, the second takes us deeper into the "forbidden" realm, beyond the forum-governing principles of freedom of speech. To make this a bit clearer let us consider, briefly, some aspects of liberty and dissent—of "doing as one pleases" and "criticizing"—as they apply to the school or appear to the teaching power.

An adult may, generally, have a choice about whether to submit himself to the disciplines of the teaching power. He need not go to the university or to a professional school. He may choose to live his life without more formal schooling if he is willing to pay the price. The choice (unless, perhaps, he is in the army or in a prison) is his. The child has no such choice. He may be compelled, for a time, to attend a public, or accredited, school; for him, to be at liberty is to be at large, to be a truant. Why do we not give him a choice? Because, although there are other reasons as well, it would be too cruel to condemn a person to a mode of life "chosen," if we can even use the term here, in a condition of innocence, ignorance, and immaturity; he is, as yet, an incompetent guardian of his own future interests. He can, to be sure, frustrate and defeat us—himself—in many ways, but at least he must report in to the teaching power.

But the principle of voluntarism reappears beyond the threshold of the school because students may feel (I believe "feel" is the accepted locution for "assert without sufficient thought") that they should have more choice about what to do or should even be governed entirely by the principle of student choice; and there are usually some teachers around, and youth-sycophant ideologues, who feel the same way.

In the world of development, however, below the age of consent, the choice or consent of the undeveloped cannot claim full sway. For mind, as for body, growth has its requirements, and what is required is not always obvious to those in need. Thus, the school has curricular responsibilities which it is not constitutionally free to abandon or to delegate unduly.[28] This is required, or this and that, or this first then that, like it or not. The teaching power must take account of liking and disliking as presenting problems and opportunities, not as limiting its authority. The student may be granted some elective options, but he must be led to whatever, in its place and season, is appropriate. It is not merely a case of formal exposure to required subjects. Habits must be formed and powers developed. The school is the kind of place where that goes on. It has a habituating mission and the necessary disciplinary power. The state, acting through its teaching power, confronts the mind in circumstances in which its authority is not defined by impulse and inclination. Here it can demand the attention and application for which, in the life of the forum, it can only plead.

The principle of liberty or of student-centered voluntarism is, nevertheless, persistently asserted. The argument, although varied, takes two main forms which I shall characterize as, first, a romantic view of pedagogy and, second, an infantile view of reality.

First, it is held that one learns (or learns best?) only when one has a desire to learn. Children, it is said, are naturally curious and this curiosity should be encouraged as the motive for learning. It is encouraged when given free rein and deadened when one is forced to learn what one is not curious about or interested in. Learning should be a self-directed form of play, and we are offered the vision of a society of addicted learners driven by unquenched curiosity, probing, examining, uprooting, creating, vanquishing ignorance, and bursting into the promised land—if only we don't interfere with the game and if we get rid of requirements and structure.

This view of learning has deep roots and has all the power of a caricature. It cannot easily be refuted, and it deserves appreciation and sympathy. Curiosity is important; enjoyment does attend learning. One would have to be a fool or worse to deliberately strip education of their support. But, but, but . . . It is simply not the case that we learn only when we want to learn or that curiosity and cognitive pleasure are sufficient to guide and sustain us. There are some, no doubt, for whom knowledge is an end in itself, in whom curiosity or a desire to learn is the master passion. But for most, and I do not say this with regret or in derogation, learning is simply a part of life. We learn in the process of doing, developing, making, failing, experiencing, judging. We learn in stride, as we cope with situations in which we find ourselves or in which we are placed. Curiosity flashes on and off, opens up or diverts; enjoyment comes and goes, encouraging, rewarding, deserting, betraying. No, it is not the aim of the school to turn us into cognitive hedonists living to satisfy the demands of curiosity; the love of knowledge is not quite the love of wisdom. Curiosity can be an asset; but it does not deserve autonomy. The enjoyment of learning is to be encouraged; but it cannot determine or govern the curriculum. The teaching power must utilize these forces; it cannot abdicate to them. The conception of the school as an autonomous playground is a disaster.

Second, student-centered voluntarism is sometimes defended simply in the name of the autonomous child. The child, it is said, is a person, and his rights and dignity as a person should be respected. He has beliefs, desires, and needs and knows himself better than others know him. He is, of course, weak and dependent, but that does not justify overriding his beliefs, ignoring his desires, or deciding for him what he needs—subjecting him, in short, to a tyrannical regime, denying him his proper liberty.[29]

I cannot undertake to present fairly or adequately the variety of views which develop this theme. Sometimes the child-adult dichotomy is rejected in toto; sometimes it is accepted, guardedly, and the dispute is over where to draw the line. Sometimes it is granted that parents may, for a time, stand *in loco parentis,* but that no one else may. In some versions skepticism and relativism are pushed to the point of denying the parental claim to know better. In others, parents and the adult society are said to know worse. The child and the culture of the young may be seen as the embodiment of the virtues—innocence, goodness, spontaneity, honesty, generosity, love—which are corroded and corrupted by death-enamoured adult culture. In many variations, "leave them alone and they will save us" is the underlying theme.

In spite of all this charming (in small doses) childishness, society, all unregenerate, declines to regard the child as the tribunal to which it submits for judgment, or even as a proper claimant to an equal voice. It asserts over its children a measure of control which it does not claim over adults. It comforts itself, in doing so, with J. S. Mill's observation in *On Liberty* that a society has only itself to blame "if it lets any considerable number of its members grow up mere children" since it has, he adds, "the whole period of childhood and nonage in which to try whether it could make them capable of rational conduct in life." So we interfere with the liberty of children and impose our culture upon them—our language and arts, our sciences and crafts, our categories and creeds—preparing them for the processes of adult life and the rights, the liberties, and the dignities of *that* condition.

Is it not obvious, in this controversy, who has the deeper regard for the person who is, as yet, a child and under tutelage? The teacher, it seems, is torn between the role of nanny and the role of guardian. As nanny, one is allied with the child against the world—comforts, soothes, shields, indulges, interposes—and lets him play. As guardian, one scans the generation with a recruiter's eye, aware of the world's tasks, seeking to fit talents to roles and to harness and realize potentialities. The teaching power, in the end, is more than nanny. It cannot be completely child centered or bound by a claim of the right—not yet inherited—to do as one pleases.

A glance at "criticism" also reveals significant differences between life under the teaching power and life in the adult forum. Central to our view of the normal political process is the conception of a stream of criticism playing heavily and relentlessly over all that we are and do. It is, we believe, essential; it reveals our problems and moves us to improvement in a continuous process. Criticism as a way of life is seen as the cultural alternative to a life of dogmatic slumber punctuated by nightmare.

The principles of freedom of speech in the forum are, in good part, designed to encourage and protect the dissenter and critic. But, in many ways, the forum presupposes the school; it assumes and needs a general condition of forum-worthiness, the ingrained habits of discussion, disagreement, cooperation. In short, the institutions of criticism rest upon the art of criticism. We expect—demand—that the school prepare us for the forum. It is not enough to turn out acquiescent schoolboy patriots; we want a constant supply of fresh, critical minds. If we are to live with "caveat emptor" in the marketplace of ideas, we must do our best, in the school, to make ourselves capable of rational—critical—conduct.

The teaching power must, therefore, approach criticism as an art to be cultivated. It must understand criticism. It must, to begin with, understand that it is closer to appreciation than to hostility. To criticize is not simply—although long experience with the "critical essay" of students fresh from high school is sobering—to "find something wrong with" or "say something bad about"; it is to exercise intelligent judgment. It is easy to find something bad to say about a book, a person, an institution, a society; but to allow that to pass as "being

critical" is to confuse hostility with understanding—to confuse, as we tend so easily to do, the loud expression of innocent (ignorant) hostility with the announcement of the arrival of the new age of critical consciousness.

Criticism is more difficult. We expect a music critic to understand music, a literary critic to understand literature.
Must not a critic of society understand something? Is every
ignorant carper to be dignified into social critic? Significant
criticism is a form of appreciation; appreciation requires understanding. The teaching power, therefore, as it seeks to cultivate critical minds, does not merely encourage and protect
irreverent outspokenness. It prods the impulsive mind into
the discipline of understanding, into deeper comprehension,
into sympathy, objectivity, fairness.

But the critical art may require more, even, than percep-
tiveness and understanding. Just as there is something strange
about an art critic who does not love art, so there is something strange—and even fraudulent—about the critic of society, of politics, of government, who does not love the object
of his critical attention. Burke says somewhere that one must
approach the flaws in his society as one would approach the
wounds of a parent. It may be difficult, these days, to know
how to do either, but what is required are understanding and
love.

The task of the teaching power is, with respect to criticism,
not an easy one. It cannot simply supply rashness with a few
tricks and pride itself on the critics it then unleashes and
sends into the world. It must cultivate carefully, presiding
over a process of growth that has its own seasons. It must take
account of timeliness, of due course, of the stages out of which
critical intelligence ultimately emerges.

Consider, for example, the relation of habits to questions.
Early education is largely the formation of habits, and questioning is one habit among others. To properly acquire the

questioning habit is to learn when and how to question and when and how not to question; it is not simply to increase the proportion of our sentences ending in question marks. Questions can be premature or belated, relevant or irrelevant, superficial or profound, helpful or destructive, pointed or distracting, proper or improper. Questioning which can be an aggressive verbal habit must be developed into a deeper irenic art. Socrates is the patron saint of questions. He stands for the examined life. Not for indiscriminate questioning, not for eristic games, but for the right question at the right time and in the right way. And he held, it should be remembered, that virtue must be habitual before it is to be questioned or criticized. Thus, when the teaching power addresses itself to the task of developing the art of questioning, it may appear, to the unenlightened, that it is engaged in taming questioners.

The well-tempered questioning attitude is haunted by skepticism, cynicism, and iconoclasm, and the teaching power is badly vexed by these seductive spirits. It is, as has been stressed, concerned with initiation into the life of the community. It seeks participation and commitment to processes and ends which must be characterized normatively. But where there are methods there are always botchings; where there are goals there are failures; where there is faith and trust there is betrayal. The introduction to shoulds, oughts, and goods is also an introduction to evil. This can be bewildering and complex where the need may be for something clear and simple. So we evoke—every culture evokes—its special parables and myths, its exemplary world. Here purposes are clearer and purer, devotion more unwavering, arts more potent than in the more prosaic world. And we begin the inevitable shuffling between two worlds—the ideal and the actual, the immaculate and the soiled, archetype and copy, normative and descriptive.

This aspect of education is, I believe, an inevitable feature of growth, and it is full of hazards for everyone. Ideals can suddenly seem illusions and the disillusioned idealist falls easy prey to self-destructive cynicism. Myths and parables can be foolishly rejected as lies instead of being cherished as perpetual invitations to interpretation and reinterpretation, lifelong touchstones. There is a need for enchantment; without it, nothing much happens. But it is shadowed by disenchantment. Or rather, there is a rhythm of enchantment, disenchantment, and enlightenment, which education must respect. There is a time for myth and a time to be literal, a time to accept and a time to examine, a time to be soft and a time to be tough. Confusion in timing can be disastrous, and even what is timely can be misunderstood. A quick glimpse of enchanting—Ah! Brainwashing, indoctrination; a glimpse at disenchanting—Ah! Subversion. A cross section of a process may not be very illuminating. But the confused outsider is more than matched by the confused insider—the ever-present educator-idiot who shouts "It's up his sleeve" in the middle of the act; the professional iconoclast who doesn't understand icons or, for that matter, truth; the teacher who applauds when the child says "The king is naked," not realizing that a "king" is never naked and that it takes imagination, understanding, and discipline to see the otherwise invisible but real clothing—not the child's half-opened eye. Problems such as this, alien to the public forum, are native to the teaching power.

A last comment on the cultivation of criticism. The critic, we say, must be independent minded, with courage to stand alone, with confidence in his own judgment. He must speak his mind, like Abdiel, "unshaken, unseduced, unterrified" by any serpentine chorus. But even here vices lurk in the shadow of virtue, waiting to pounce. Independence, yes. But not the incipient idiocy of the loner, stubborn, hard-core adamancy.

Self-confidence, perhaps, if warranted. But self-esteem, pride, is still, although this seems hard to remember, a terrible vice. In the healthy critic, independence must be tempered by modesty and humility, by the awareness of other minds, by the occasional recollection that the common herd contains one more member than each of us supposes. The critical independence we seek to cultivate is that of a partner, not of a crank.

This cursory glance at the shape of liberty and criticism in the domain of the teaching power should remind us that the school cannot be essentially understood as marketplace or forum and that the teaching power must wield routinely powers which are frightening in scope and implication and which are, perhaps, uniquely its own. The legislature, we say, cannot legislate morality. Can that be said of the teaching power? Not if we understand development in the context of initiation. The school seems inevitably to moralize. It is a dangerous place. We are concerned, therefore, to control the teaching power and prevent its misuse. We embed the school in a context of political controls and constitutional constraints. We insist on due process. We may encourage competitive and alternative schools. But this is not enough. We can hardly add—although some are tempted to—the principle that the school should leave the mind alone. But we do attempt to make distinctions and develop principles that clarify the proper function of the teaching power and which, if observed, would keep us safe. Some of these principles are better than others; some are misleading; all, I am sure, invite interpretation and require that we reflect upon important questions of educational and political theory. I turn now to a brief consideration of some principles proffered in the hope of fending off threats to freedom posed by the undeniable presence of government, in the guise of the teaching power, in the realm of the intellect and spirit.

That the School Should Stick To Facts

Threatened by controversy it may seem wise for the school to retreat and take its stand on information. Here is the community torn by conflict over race, religion, sex, and politics. What is the school to do? It cannot altogether avoid everything that is controversial. But can it not, at least, eschew judgment and confine itself to dealing in information and to sticking to the hard facts? Let the school teach, it is said, the facts about religion, the facts about sex, the facts about race, the facts about capitalism and communism, the facts about history. Then the student can make up his own mind, or someone else can make it up for him—parents, priests, scoutmasters—as long as it is not the school. Education then takes its stand on the acquiring of knowledge, on information, on getting the facts. Facts may be presented or, better still, the student can be taught to dig them out for himself. He is taught to demand them, to respect them, to gaze at them unflinchingly, to accept their verdict.

Respect for facts is not a negligible virtue and I do not mean to disparage it when I suggest that it is not central in the educational drama. It supports, but it does not lead. The facts do not present themselves to be served up in passive heaps, nor in complete collections, nor stripped bare, nor underlined for significance. They simply cannot be doled out, or gathered and assimilated *first,* without guidance by nonfactual considerations.

Facts are digestible or significant only in context. When we are entertaining a belief, an hypothesis, a theory, then facts come into play. They are relevant or beside the point, they support or shake, they are insufficient or decisive, convenient or inconvenient, they help explain or need to be explained away. But the significant context is more than a context of belief. It is a context of actions, of enterprises, of purposes that, in turn, lend significance to belief and theory. In this broader practical context, facts, as they bear on action, can be upsetting, discouraging, destructive—one man's triumph, another's defeat.

To the teaching power, "just stick to the facts," is advice that is easier to give than to follow. It cannot really strip education of its context of significance or values; it cannot organize itself around the bare facts. There are, for example, facts about sex. What is to be done with them? Are they to be dumped helter-skelter before schoolchildren? Or are they to be placed carefully in a context of love, family, society, cosmos? Clearly, facts must be handled with tact. (If the sex example doesn't move you, try race.) They must be placed in context, seen in context, understood. There are educational tactics about facts—conditions of readiness which must be respected, questions of emphasis and focus, preparation and postponement, discretion and revelation. Facts are to come in time and season, not in an indiscriminate flood. But the flood is controlled by policy, and policy is controversial, and we are back, almost, to where we began. We cannot find permanent educational peace under the aegis of fact.

Nor is it clear that, if we could, we should. The teaching power, I have argued, is concerned with development in the context of initiation, with induction into ways of life, into modes of action which are purposive and value laden. Do we really mean—separating "fact" and "value"—that the schools should inform about honesty but not cultivate it? Inform about race but not affect racism? Here are the quests for truth, justice, beauty—take it or leave it. This is proper usage, should you ever care to use it. Are we running a gigantic department store, displaying all the options, teachers attending every counter, careful, in the name of free choice, not to influence the customers? Utter nonsense! The school is not a store. Students are not customers.

The relation of fact to context and the necessities of initiation defeat simplistic attempts to erect the fact-value distinction—itself rather tattered, by the way—into a basis for the limiting of the authority of the teaching power. The school cannot just stick to the facts.

That the School Should Just Teach "Methods"

The school, if it cannot simply stick to the facts should, it is said, teach not *what* to think but *how* to think. Learning is seen as learning how, as acquiring skills—reading, writing, calculating, arguing, proving, deciding. There are methods of inquiry, of disputation, of decision-making—supported by methodologies. The school is to transform these methods into habits; the method of inquiry, understood by the teacher, becomes, in due course, Johnny's habit of inquiring. Since we do not know, and therefore cannot teach him what he will need to know, we teach him how to find out; since we cannot teach him what to do we must teach him how to decide for himself. The school, in short, is not to stuff students with information but is to train them as enquirers; it is not to supply them with opinions but is to prepare them for controversy. Not "what," but "how" is all.

The power and appeal of this educational methodism is obvious. First, it celebrates the centrality of habit and, icon for icon, habit is better than fact as the object of educational adoration. (Mind is better seen as muscle than as bottle.) The teacher understands that skills are central and that imparting skills is forming habits, accustoming.

Second, teaching "how to" seems to respect and enhance the active independence of the student. Learning how is acquiring the ability, the power, to do something. The more we learn how, the greater our power, the greater our freedom to do or not to do. There is a real sense of liberation that one experiences as he learns how to do it himself. Who cannot recall the feeling of freedom and power that grew upon him as, learning to read, he found he could conquer libraries alone. Learning how frees us from dependence, from the transient fashions of teachers, from narrow, limiting context. We become empowered.

And finally, in teaching "method" the school seems to serve the community without becoming embroiled in its quarrels. Is there conflict over tradition and change, liberty and equality, left and right? Certainly. But the school need not decide who or what is right; it need not declare for Yin or for Yang. It tries to teach how to behave in the midst of controversy, how to solve problems and resolve disputes. It is everyone's coach.

This attractive view of education seems to promise the avoidance of partisanship through a deeper commitment to methodology. It deserves respect. But, of course, it has its problems. To begin with, faith in "method" may be a misplaced faith. There are some areas in which we speak with confidence—"the scientific method," for example; but even here, while there are zealots who regard its teaching as the panacea for all ills, there is an undercurrent of professional skepticism which provides a derisive undertone to naïve enthusiasm. And when we get beyond the hard and triumphant sciences and look around for methods of decision-making, conflict-resolving, of dealing with moral or value questions, the faith in methodology gets stretched to the breaking point. To teach methods in these areas seems less to teach how problems are to be solved than to teach how to blunder, stumble, cope, suffer, and endure. There is, at any rate, some allowable doubt about the efficacy of faith in method. It sometimes seems that the farther one gets from the genuinely creative or productive master-worker the more one hears eulogies to method.

Nevertheless, teaching is largely teaching how. But teaching "how" is more fundamentally teaching "how we . . ." rather than "how to . . .". Teaching a child to speak is, after all, teaching him to speak as we do, with some marginal play between how we do and how we should speak. Teaching how is always teaching something in particular. That is, we teach (and learn) English, not language; tennis, not game; democratic politics, not governance. In learning something general we learn something particular first.

Teaching "how" it turns out, is initiating into a particular, active fellowship. Someone is to get involved in what is going on; he is to get into the habit of doing something, to share in, participate in an institution. The habits, the methods, the ways, which the teaching power cultivates in its special environment prefigure, reflect, and prepare for the institutions, the culture of which the teaching power is the agent. The more we learn "how" the more we become involved, the more fully our habits incorporate our institutions, the more inevitably our character exemplifies our culture, the more deeply we become part of the going system.

And there, for some, is the rub. It is difficult, if not impossible, to habituate to what is not institutionalized, to teach a habit not supported by the environment, to shape a utopian character in an imperfect world. The Tao is the way; the way is the method; the method is the habit; the habit is wedded to the established fact. To those who do not realize that all our trips begin from where we are, it all sounds too much like a conservative plot.

That the School Should Be Neutral

Besides being urged to stick to the facts and to teach how, not what, the teaching power, in order to serve the whole community, is urged or required to embrace neutrality as a guiding principle. This is a valid enough principle where it is applicable, but its scope is narrower than is often supposed and it can, when its limitation is not realized, cause serious confusion.

Neutrality is called for in the face of the partisanship of others; one is neutral as between combatants. I stress "as between." Neutral "against" is a familiar joke. Neutral "about" is a locution suggesting indifference, "not caring," lukewarmness rather than impartiality; and that is not the sense in which neutrality is recommended to the teaching power. Neutral *as between* . . . and, moreover, as between alternatives *on the same level*. That is, a referee is called on to be neutral between Ohio State and Michigan, not between Ohio State and Football. Or, to take another example, a judge is to be neutral as between the defense and the prosecution in a case before him. That does not mean that he is neutral in any sense about law. He is not—it hardly makes sense to talk this way—neutral as between the accused and lawfulness. He is not required to be lukewarm or indifferent about, or uncommitted to, legality.

The familiar American educational example has to do with religion. "No establishment" at least requires, almost everyone agrees, neutrality as between religions in the public schools. There is to be no favoring of Protestant over Catholic or Jew. But this, of course, should not be taken as implying that the school must be indifferent to religion or even that it is to be neutral as between parties to *another* dispute between the religious and unreligious views of life. Perhaps it should be, but that is another question and it should not be settled by momentum borrowed from the requirement of neutrality as between religious sects. We have drifted into such confusion about the implications of neutrality for education and religion as to evoke from Justice Douglas the remark: "We are a religious people whose institutions presuppose a Supreme Being"—a heresy for which, as a reward for his many other deeds, he may be forgiven.

Much the same problem appears in connection with the proper demand for political non-partisanship. When we say that the school should stay out of politics, we intend that its

operation should not reflect partisan bias. It is not for this
party or that, for Republican or Democratic; it is neutral as
between parties. But this does not mean, as it is sometimes
taken to mean, that the school is to be indifferent about politics in general or about the political system. Non-partisanship
serves a deeper commitment to politics. The school is, among
other things, to prepare people for political life, to develop
the capacity, the understanding, the attitudes necessary for
the operation of the society's political institutions. It is indifferent to the fate of this or that party, but it is deeply concerned with the health of the deep political or constitutional
structure. It educates members, agents, critics; it furnishes
the establishment; it supports the system. This is not an ac-
cusation; it is a statement of its function.

The danger in the careless use of notions of neutrality and
non-partisanship is that the concern for fairness may be taken
as requiring the relinquishing of commitment. Or, on the
other hand, the obvious commitments of the school may be
taken as reducing any claims of fairness to mere pretense and
hypocrisy. These are foolish views and result, I repeat, from
confusion; but this is a point around which fools, impatient
with distinctions, seem to swarm and buzz. "Neutral" and
"non-partisan" are necessary conceptions. There is a narrow
sense of these terms suitably applied in limited contexts. But
if we take them out of these limited contexts and attempt to
apply them more broadly, we do not adequately express the
relation of the teaching power to the community and, in confusion about fairness, become confused about commitment as
well. "Neutral" does not express the fundamental relation of
the school to the society it serves.

Seeing the teaching power as an inherent power of the polity we have considered ways in which its frightening power

might be checked. Decentralization, the proliferation of private and parochial alternatives to government schools, attempts to define and limit the teaching role may, together, be quite effective. But there is something chilling about the very conception of the teaching power itself, about the conception of the teacher as the agent of the state. Can we, somehow, get out of *that?* I do not think so. We can ignore it or pretend it isn't so, but there is no salvation in the averted gaze. Can we defiantly place the teaching power elsewhere? In the hands of the (a) Church? The Family? Can we simply renounce or relinquish it? No such mere institutional moves will solve problems any more effectively than will the principles of separation and distribution of powers, of decentralization and delegation of authority, within the state. The teaching power inheres in the state as clearly and inevitably as does the judicial power.

Can we, in that case, at least formulate the function of the teaching power in such a way as to keep the agent from servility, to keep the school from becoming the instrument of the sins of the fathers, to aid the teacher in an attempt to shield the generation in its care from the influence of a sick or evil social order?

As we approach this question let us note that the same range of problems exists also in the comparable case of the judicial power. The judiciary practices its art within the structure of the judicial office. It is, clearly, a branch of government. Judges are officials, agents of the state. Their function can be described in different ways, but two things are centrally involved. First, judges decide certain disputes in a context of, with reference to, and in accordance with the positive law of the particular society. This law varies from society to society and is subject to judgment as foolish, harsh, or unenlightened and worse. Nevertheless, the positive law has a strong initial claim to judicial deference whatever the private views of the judge may be. It is his job to apply it, to interpret it, to say what it requires. He is sometimes called the servant of the law. But second, the judicial concern is also for "justice," and in its name, in many guises, the judge subjects the positive law to some higher-law influence, checking, frustrating, mitigating, reinterpreting the apparent will of the law-maker. It would be a naïve mistake to dismiss justice or constitutionalism, or the higher law, as a mere "brooding omnipresence in the sky." It is a fundamental feature of the legal order which keeps the service of legality distinct from mere subservience. Popular judgment, with a healthy instinct, has little difficulty with the dual conception of the judicial function as "enforcing the law" and "promoting justice," is perfectly aware of the possible divergence between law and justice, and accepts "judicial independence" as necessary if the judiciary is to be able to do its job properly. It does not occur to anyone that because a law may be unjust our judges, in order to promote justice, should not be agents of the state or public officials.

I find the parallel between the judicial and the teaching power very striking in many respects, but the similarity relevant here is this: just as the judicial power has a commitment to the "given" in the form of the society's positive law, so the teaching power has a commitment to the given expressed as the function of initiation into the ongoing society. And just as the judicial power cherishes and insists on a commitment to a normative higher-law or justice, so the teaching power in the midst of its mundane initiatory tasks insists that it must also keep its commitment to the logos, or rationality, or virtue, or objectivity, or perhaps, the free mind—some transcendent, culture-critical higher-law ideal. I will consider this in a moment, but let me note again that to preserve the "higher-law" aspect of the teaching power's function it is not necessary, anymore than in the case of the judiciary, to free the teacher from his role as agent of the state. Academic independence—the separation of powers—is enough.

What, then, do we want when we are dissatisfied with a merely initiatory role for the teaching power? What do we fear? We want the teaching power to protect our children from our errors, our follies, our vices, our shortcomings. We fear that a teaching power too timid, too uncritical, too responsive, will only perpetuate our betrayals of the ideal. And we fear, in the extreme case, that, if and when the state falls victim to tyranny, to the dictatorship of the zealot, a docile teaching power will dutifully warp the mind to its decrees.

In the context of these anxieties it is understandable that there is reluctance to accept the commitment of the teaching power to initiation into what is going on, and that there are recurring attempts to get around that commitment. One attempt, of course, is to deny that initiation is, in any sense, part of the teaching power's function. Stress is placed on helping the learner develop himself as he sees fit and leaving him free, so to speak, to attach his energies to stars of his own choosing. I will not linger over this plausible view which produces, from time to time, child-centered schools with administrators who see their tasks as pandering to kids and keeping the adult culture off their backs. If, at this stage of the game, it is necessary to explain what is wrong with this to anyone, he is probably incapable of understanding the answer.

But the other attempt is more interesting. It accepts initiation as an educational necessity but claims, in a utopian spirit, the mission of initiating into a higher order of life than is exemplified by the imperfect existing culture. It attempts to shape an ideal character fit for an ideal culture, for the brotherhood of man, for a rational cosmopolis, for a "perfect" democracy, for the good life—whatever that may be. In this attempt it may find the existing culture of the community an obstacle to be swept aside or surmounted and it may come to regard the alienation of the student from *that* as a preliminary triumph.

The elevated tone of all this is, no doubt, appealing. It captivates the idealistic teacher disillusioned with a society that espouses competition for material goods—and all that. But I cannot really end on a note of acquiescence in a self-righteous declaration of independence by the teaching power in the name of its service to a higher morality. First, there is always the question to be put to the self-appointed guardians of the higher goods: Who are you? Where did you get your special insights? On what road did you find illumination? The local School of Education? The Teachers' Union? A meeting of the committee on the curriculum? Who appointed you to veto the culture?

But, second, there are profound difficulties in the path of any attempt to derive from the contemplation of an "ideal" a specific prescription for conduct. Thus, we cannot derive from the idea of justice a proper or ideal code of positive law; we cannot derive from the idea of a good society any particular set of social institutions; we cannot derive from an understanding of rationality the particular beliefs we ought to hold. We cannot, to add an odder example, determine from the most profound understanding of "language" what language the ideally educated person should speak, nor create the ideal language to supplant the poor excuses for languages that people happen to use. The higher law, the higher ideal, is critical, not creative; it may improve a way of life, but it cannot bypass it or provide another in its place. The road to paradise lies through and not around the here and now; if we are to get there, we must get there from here.

Teaching-power utopianism ignores these truisms and produces, from time to time, an other-worldly school with its strangely victimized students, or a gaggle of counter-culture gurus leading sullen children in circles, but on the whole the teaching power is saved, perhaps by lack of imagination, from messianic delusions. Its basic style is more Sancho Panza than Quixote, more Sam Weller than Mr. Pickwick, more Jeeves than Bertie Wooster. Its strength is in its pedestrian sanity and if it seems, at times, to lead its master a bit, it leads, nevertheless, in the master's direction, in his terms, under his banners. It teaches English, if that is the going tongue, as she is spoke, perhaps, a bit, as she should be. And it does not harbor the illusion that all the errors made, lamentably, in English, would disappear if only we could be induced to speak in French.

Keeping the public school in the center of the stage as, in principle, undeniably legitimate, I have argued that it is an institutional expression of the teaching power, understood as a fundamental constitutional power of the state. The scope of the teaching power is enormous and the danger in its exercise, no less than the dangers of leaving it unexercised, requires the constant attempt to clarify and safeguard its proper goals. We need to see the school as a primary agency of the state exercising its authority over those who are still below the threshold of consent; we need to understand the special powers implied by the related tasks of development and initiation and the very limited assurance against misuse provided by over-simple theoretical attempts to disengage the school from politics. Education must be understood as an enterprise that places government fully within the sphere of the intellect and spirit and that requires, as a part of pedagogic theory, an understanding of the political aspects of the habitual structure of the mind.

4

Government and the Forum

"We govern speech in order that we may govern ourselves by speech."

Burke remarks that a constitution is something to be enjoyed, not discussed. I suppose he means also that what has worked itself into our bones supports us efficaciously without the aid of understanding. This insight is, I believe, strikingly exemplified by the intricately developed constitutional structure of freedom of speech or communication in America. It is one of the wonders of the world. Any description of the American way of life must begin, not with our material prodigality nor our rough energy, but with a bemused contemplation of the communication flood which seems to sweep everything before it, penetrating everywhere, unconfined by barriers of order, of reverence, or reticence, or decency, or shame, or confidentiality, until the deepest of mysteries became mere open secrets and nothing escapes unscathed—nothing. And yet we seem to survive—I would say thrive, but I don't want to stop to quarrel about it—not in spite of but almost because of this careless willingness to live on the edge of public exposure. It seems excessive. Obscenity loses its edge, private tragedy sobs into a million living rooms, subtle diplomacy is shrilly derailed, the betrayal of confidence becomes a national sport. Scandalous! Indecent! Impossible! And yet, much as we may to want to stop it, when

86

we catch our second thoughts we don't want to stop it after all. Better to endure than to censor, we think, better to grow inured. So we nod and solemnly echo "free speech . . . first amendment . . . , clear and present danger . . . , no prior restraint . . . , no big brother . . . ," as if we understand what we're saying. It is, however, rather difficult to explain to a foreigner who may want to take it home with him. The system which works so well for us seems, like a good local wine, not to travel very well. It is full of quirks and idiosyncrasies that can be accounted for if we can remember the cases, but it is very hard to describe simply and coherently and, I think, we have given up trying. What did Burke say? Ah, yes. Relax and enjoy it.

I do not intend to describe or explain the American system of communication freedom. But I do intend to pursue some questions which will reveal the basic issues of communication government with which our system, or any system, must grapple, and which our complex practical success tends to obscure. I hope to show that government is more deeply involved in this aspect of the life of the mind than we may think, that its vigorous intervention is something we frequently demand, and that "free speech" is not what you have when government leaves things alone. What I offer here is a kind of theoretical free-speech primer.

Speaking (communication, written or oral) is an especially consequential form of human action. By speech we enlighten or confuse, lead or mislead, wound or console, amuse or enrage, unite or divide, create or destroy a community. It is an art whose every use is shadowed by a possible misuse. As a form of action, speaking is continuous with other actions which fall, unquestionably, within the proper control of the legal and political order.

But an essential feature of speech is that it is the expression of a mind. A speech act is a mental act. The attempt to govern actions by governing speech must, therefore, come to terms with the fact that speech is the means by which a mind seeks expression—not only self-expression as a private act, but communion and communication with others in the course of a common life in a public world. Saying is both thinking and doing.

Thought itself is heavily dependent on the institutions of expression. Freedom of thought and freedom of expression appear, to John Stuart Mill, "practically inseparable." Certainly, the freedom to think as one pleases without freedom to hear and to say is a hollow version of the freedom to believe. But the saying, which is so essential to thinking, is a kind of public doing; it is action which has consequences for others. Why, among action that may harm, should it escape the social curb?

It is this dual status of communication, involved with thought and action, and the dual concerns, to protect the freedom of thought and to prevent harmful actions, that complicates the attempt to understand and order Babel. To touch the word is to touch the mind, so we may not touch the word. On the other hand, we may control what may produce harm; the word may produce harm, so we may control it. Without a special concern for the mind the governing of communication would be simply and naturally the governing of a consequential mode of action.

I shall not linger here over the familiar arguments about freedom of speech—Milton, Mill, Holmes, and all that. I will assume that we share this great living tradition, that we believe in "freedom of speech," and that we need, rather, to have some of the background sketched in.

Speech may seem, to an uncomplicated view, a generally aggressive form of action, almost, by its very nature, a form of assault. The range of offenses runs from nagging through heresy and treason, and the range of punishment from the stocks to death and damnation. Verbal aggression is not naturally licensed or privileged but falls, rather, into an historically familiar catalogue of crimes.

First, verbal assault (oral or written) against public authority is, traditionally, a serious offense. At many times, in many places, the divinity that hedges the king, the respect, the confidence, the awe even, in which our ruling persons or institutions are held may be all that stands between ordered life and chaos. This is, no doubt, incomprehensible to those who regard reverence or respect for authority as something that stands between us and real freedom, but with an effort of imagination we may, perhaps, glimpse something of that archaic, un-American point of view. A touch of Yeats perhaps . . .

> Civilization is hooped together, brought
> Under a rule, under the semblance of peace
> By manifold illusion; but man's life is thought,
> And he, despite his terror, cannot cease
> Ravening, raging, and uprooting that he may come
> Into the desolation of reality . . .

Imagine a community barely emerging from a divisive ordeal, factional civil strife, and making its way precariously toward civility. Respect for government, monarchic or democratic, respect for law in its mysterious sovereign majesty, respect for the constitution, may be its only hope. Respect is the alternative to violence; and we may wish to avoid the desolation of reality, to check the ravening, to preserve the illusion, if it is that, that hoops us together under a semblance of peace—the illusion of majesty, or, in these prosaic days, of integrity or trustworthiness. But, I suppose, we have little respect for awe and respect. Our style is irreverence and the consequence is cynicism and a public life that is a perpetual crisis of confidence, with violence close to the surface. "No respect!" I do not pause to argue with those spiritual bookkeepers who protest that they believe in respect provided that it is "earned"—missing, of course, the whole point.

At any rate, in the world generally, awe, fear, respect have long histories of association with political authority, and the basis for that association is not altogether trivial or fraudulent. Verbal assault which threatens authority by stripping it of its aura has seemed to threaten the society itself and thus to merit control and punishment—whether as *lèse majesté*, as seditious libel, as anti-state or counter-revolutionary activity, or as unpatriotic subversion. Truth, it should be noted, may be quite beside the point, for it is believed, and it may be possible, that the harm done to the community by discrediting authority may exceed the harm done by its unpublicized vice.[30]

One's feel for this problem is a function of many things— one's understanding of authority, one's attitude toward reverence and its kindred humility-breeding sentiments, one's tolerance for mystery and ambiguity, one's faith in manners, ceremony, ritual, one's experience with what lies on the other side or beneath the veneer of civility. It is nothing I care to argue about with muckraking realists. It is a fact, however, that the verbal assault on public authority is still regarded in most places in the world as, in some degree, a crime. The shock it generates testifies to a deep social instinct.

Second, and equally universal, is the view that religion needs special protection against verbal assault, against blasphemy. Or, perhaps, that society needs its religion so badly that it must protect it from verbal disrespect. The sacred and the holy demand a hushed voice, a special intonation, an immunity from verbal attack.

Here, too, Americans, brought up on the separation of church and state and the consequent displacement of blasphemy as a legal offense, may need a special effort of imagination to recapture the sense of urgency that others might have about this question—people who feel that religion or God is the keystone of the entire edifice of values, virtues, of conceptions of morality and the good life, of civilization itself, and who do not take the verbal disparaging of the foundation of life as a harmless manifestation of social eccentricity. Where, in the modern world, the boundary between religion and politics gets a bit blurred—as with some totalitarian dictatorships—questioning or criticism, greeted as verbal assault, groans under the combined weight of both seditious libel and blasphemy.

Third, related to blasphemy but more removed from direct connection with formal religion is a baffling class of speech offenses which do not fall familiarly under a single rubric. They can be thought of as verbal violations of social taboos, although that characterization may not suggest adequately the deep values involved. Sex, for example, is not a trivial matter and a society may be very sensitive about it. Not out of fear, or stupidity, or neurosis, but out of a healthy respect for a basic human passion. Obscenity is a puzzling notion, but we can begin to understand it when we see that it is a kind of blasphemy about a central value. A community may insist on a special public language, a special reticence, a veiled obscurity, "decency," and it senses—confusedly but doggedly—that irreverence at this point is deeply destructive. For this unerring perception society is called "up-tight" by those who are not yet spiritually toilet-trained.

But sex is only one example. Consider ethnic names. People who are quite free in violating verbal sexual propriety may, in turn, be extremely offended by the use of disrespectful racial terms and quite eager to insist on censorship or punishment for this form of obscenity. It is not merely a question of preventing fights or riots; it is a matter of dignity.

Communication, in short, is hedged by taboos that protect values and sensibilities. A community may insist on good manners, on respect for its verbal or symbolic conventions, and it may even resort to law to control this form of verbal aggression. Obscenity is the clearest example, but it is, perhaps, too narrow to cover the entire range of verbal taboos. This is, I think, a surprisingly neglected area of free speech theory.[31]

Finally, there is a traditional class of speech offenses against persons—libel and slander—which are regarded as punishable even in societies that regard themselves as enjoying a full measure of freedom of speech. The individual, it is thought, may be protected in his reputation, his standing, his livelihood, against certain forms of attack, and there are some attempts to extend this protection to groups.

This is nothing like an exhaustive catalogue of speech crimes; but is, rather, a classification of various forms of aggression which societies generally regard as consequential, as subject to social control even though the action is merely verbal. The social control may be exercised through other powers than the legislative and judicial; it is obvious that the teaching power has a significant role here. Teaching "how to speak" a language is teaching how to behave linguistically, how to express oneself properly about different things in different contexts. Knowing or having the "manners" of speaking is a part of knowing how to speak, and the schools have this initiatory task even beyond the mere avoidance of communicative illegality. We may well regard education rather than legislation as the proper approach to this problem and it is interesting to see the school, in its current disarray, fumbling the task, hesitant about making claims against urchin argot or subculture jargon and even accepting "legality" as the only criterion—and a shaky one at that—of proper usage.

If I begin with mention of verbal aggression, pervasive

speech crimes, and universal attempts at social control, it is because I believe it is better to start with the sense of speech as an activity seen from the beginning as laden with consequences and tightly bound by customary and legal restrictions than to start the story with some supposed natural right of self-expression before which every proposed constraint must plead its cause.[32] Freedom of speech lies at the end of a long social road, not in the state of nature at the beginning. We must see it not as a primitive right but rather as something wrested from the domain of control; its history is not that of the loss of right but of the winning of privilege.

The story, if it were written, would trace the development of our understanding of the need to provide special protection or "privilege" to those engaged in the dangerous life of communication. It might begin with the emergence in the dim past—already an established institution in Homer and so familiar as not even to be explained—of the herald, the divinely protected messenger moving between armies with the ritual of parley, so that minds may meet and transform the clash of arms into truce, into dialogue, even into agreement. Blessed are the heralds of peace!

We might linger briefly over the sad fate of the hapless bearer of bad news to the short-tempered king. "For diligently bringing me the news of the defeat of my army, this bag of gold; for daring to spoil my day, off with your head." Eventually, if we want the news, we must master the distinction between the medium and the message and grant immunity to the messenger.

Beyond this lie the deeper hazards of advising, predicting, truth-telling.[33] The institute at Delphi supplemented divine protection by developing modes of oracular ambiguity as protection against the charge of bad advice. ("You should have asked *which* kingdom would fall.") Prophetic privilege is purchased through special disability and at a heavy price.

Teresias is blind; Cassandra must always sound implausible. Or the court's plain-speaker comes disguised as a fool or jester, tolerated as entertainer or madman; but he remains while normal Kent is banished. Marked as under special protection, immunized, or privileged, is the emerging conception hedging the adviser.

The more conventional history comes to focus in seventeenth-century England with the struggle of Parliament for the power to discuss and advise as it sees fit, with immunity against being called on royal carpets, against having "to answer in another place" for what is said in the House. With the winning of this parliamentary privilege (adopted as legislative privilege in the American Constitution) we are well on the way to the daring conception of an official loyal opposition (elsewhere, a devil's advocate). It is ironic that the House of Commons has little sympathy for Wentworth's plea that, as an adviser to the King, he, too, should be considered protected against having "to answer in another place" for his, he believed, confidential advice. Executive privilege has a harder path.

The perception that, when the performance of a legitimate function calls for the candid expression of views, it is necessary to protect the speaker or adviser by immunizing him against certain forms of reprisal lies behind the early disappearance in America of the crime of seditious libel, as incompatible with a system in which public officials are answerable to a popular electorate. Thus, freedom of political discussion can be seen as the extension of parliamentary privilege to the electoral branch of government. The secret ballot can be understood as a device to protect the voter, in exercising his judgment, from having to answer in another place.

The full story would go beyond politics to the unleashing of discussion and criticism in almost every area of life. But "unleashing" is the wrong word. It is not simply a freeing from restraint, but the providing of immunity or protection that is crucial. It is, I think, significant that our symbol of freedom is not the call to "stand up and be counted"—an invitation to heroism—but is rather the canvas curtain behind whose shrouding protection, provided by civic artifice, we finally say, uncoerced, what we really think.

The glacial social movement into what Bagehot has called "the age of discussion" would, if appreciated, be seen not as a return to a condition of social innocence by stripping off the accretion of restraints on a natural freedom of expression; but, rather, as the gradual creation of a great system of artifact, the forum, which through its structure of rules, opportunities, and protections, gives meaning to the essentially civic conception of freedom of speech. The relation of government to the forum and to the life of mind within it is the theme of what follows.

I shall use "forum" to refer not only to a particular place where speech or communication takes place, but to designate also the whole range of institutions and situations of public communication. This usage is analogous to the way in which we speak of the "market"—referring not merely to the supermarket on the corner or the stock exchange, but to the broad range of economic transactions as well.

The forum is essentially a system of opportunities and protections—opportunities to enter into communication and protection against some of the consequences of doing so. There must be both a place of speech to which the speaker has access and some protection against having to answer in still another place. This dual requirement can be easily grasped if we consider together the traditional phrase, "no prior restraint," and its possibly disconcerting complement, "subsequent punishment." We can imagine a system of prior restraint—permission needed to meet, to speak, to publish; prior approval required by censor or licensor—so comprehensive and tight as to make subsequent punishment unimportant and rare. The absence of prior restraint, on the other hand, could be reduced to pointlessness by a heavy system of subsequent harassment and punishment. If we are to have any prior restraint—and I do not think the suggestion is always necessarily pernicious—it must be compatible with principles of reasonable and fair access; if we are to have subsequent punishment that is compatible with an adequate forum—and some is—it must be for clearly proclaimed excesses, with some measure of tolerance for the regrettable but normal forms of verbal aggression, and with supporting reassurance for the timid—or else the opportunities will be seized only by the rash and the heroic.[34] To provide a forum is to provide both opportunity and protection.

The point of the forum, apart from its inherent delightfulness, is to provide the community with the opportunity to develop its mind or consciousness, for public and private purposes, through the exchange of ideas. There must be, for any community, a forum adequate, in extent, to its purposes. "Adequate," of course, does not administer itself and becomes a political problem. It is certainly not very clear, but it is not hopeless. Assembly places must be provided, with some regard for convenience; there must be opportunities for dissemination, face to face or indirectly. And so on. The principle is simply that people must have opportunity to meet, to discuss, to speak, to hear. It is partly a problem of architecture and city planning, of the technology of communication, of the psychology of interaction (enormous squares? small parks? sidewalks? halls? amplifiers?). We can be starved of opportunity or glutted into a daze. "Enough" is an interesting problem. When do we have adequate forum?

The principle, if we can call it that, of adequate opportunity is supplemented by a principle of fair access. It would be futile to try to achieve a one-man—one-voice situation such that no one could be said to have a greater voice in the forum than anyone else. But an adequate system will attempt something in the direction of fairness and will try to provide access to channels of communication, in one way or another, for everyone. Not everyone can own a newspaper or radio or television station; nor even, if these are publicly owned or controlled, have equal use of space and time. But government may, in the interests of persons who have a claim to be heard and in the interests of the public which has a need to hear, move to provide for fairness in the struggle for access to the mind. It may establish rules for equal time or right of reply. It may enforce antimonopoly rules and even subsidize competition. Facilities can be provided at little cost on a first-come, first-served basis. At public meetings or hearings, rules of recognition, time limits, and systems of representative spokesmen can be adopted. In these and other ways we can give some substance to the principle of fair access. Where government itself operates aspects of the forum, fairness requirements are built into administrative regulation. Where private institutions operate, natural partisan tendencies and heedlessness about fairness may be mitigated by government regulatory policy. But however it is provided for and however much we may fall short, fair access is an essential aspect of forum adequacy.

If we really wish to encourage people to speak out, to say what they really think, we must supplement opportunity with protection. Perhaps the most familiar protective device—paradoxically—is confidentiality or anonymity, or even secrecy. The secret ballot, already mentioned, is designed to protect the voter from having to run a gauntlet of animosity, resentment, or threat. The closed meeting, whether of a city council, or school board or jury, is designed to encourage freedom and candor within a particular context by shielding it from the coercive pressure of publicity. The assurance of confidentiality may be the condition of getting sensitive advice and dependable letters of recommendation. Anonymity may assure free expression on questionnaires, in polls and surveys, in letters to editors, in leaflets. And so on.

When impunity cannot be provided in these ways, we may try for immunity more directly. A legislator may be privileged against prosecution for libel or slander, a witness may be offered immunity for testimony, an employee may be protected against being fired for the expression of political views. Additionally, we may nullify, under the aegis of the forum, the force of some of the traditional speech crimes; the citizen discussing public affairs may be privileged to libel with virtual impunity. The aim in all this is to avoid a "chilling effect" on the freedom of speech.

There are two aspects to the problem of protection: protection against punishment by government, and protection against private reprisal. The first has been, on the whole, our greater concern. In addition to the instinctive tendency of governments to cherish the category of speech crimes—to yearn, almost, to punish for seditious, blasphemous, obscene, libelous, disorderly, or provocative utterances—there is always the danger that long memory or inescapable dossier will be used to subject the outspoken citizen to harassment and to prejudiced consideration of his claims to opportunities and benefits. These problems, familiar and difficult as they are, may obscure the coercive dangers of private reprisal. The police can normally protect the soap-box orator, the picket, the peaceful demonstration or parade, the public or private meeting, against individual or mob violence. But government may be baffled as to how to protect individuals who express unpopular views against subtler forms of private vengeance— against the poison-pen letter and anonymous phone call, the whispering campaign, character assassination, insult and rudeness, and various forms of ostracism and boycott by which the free-speech vigilante drives the dissenter from the local consensus into wounded silence. Where legal remedies are fruitful, they may be used. But the burden in this area must, in the end, be borne by our educating and moralizing institutions. Much of the failure to sustain, encourage, and protect the forum participant is, beyond the failure of government, a failure in the morality of discourse.[35]

The forum, to sum up, is a system of opportunities and protections. It is a necessary institution, and government may share, at least, responsibility for its adequacy. Both opportunity and protection must be *provided*. And since such provision is normally by and through law, government is involved in the very fabric of the forum. The forum is not something we have before government takes a hand and which is "free" until government intrudes. It is a system in which government is a constitutive element. To say that government should, in the name of freedom of speech, leave the forum alone is like saying that government should, in the name of justice, have nothing to do with courts.

As a prelude to the consideration of problems of governing the forum in its broad sense and as a gesture of respect to one of the great archetypes let us pause to contemplate, for what it will recall to our attention, the semi-mythical council or town meeting—the exemplar of freedom of speech.

The freedom of speech, it has been well said, is not to be confused with unregulated talkativeness. So we begin with the gavel and the call to order. The silence it brings is the beginning, not the end, of communication; and from now on a speaker must be recognized by the chair and given the right to speak. The state of nature is replaced by civil society.

The call to order is itself heavily subject to regulation. Has the meeting been properly called and announced? Has due notice been given of where and when, and of its purpose or agenda? Is it sufficiently convenient so that those who have a claim to attend and participate can do so? Is it open to non-participants and reasonably adequate for audience and press?

The chair—and the meeting must be chaired—is the guardian of the agenda that is, of course, the heart of the meeting. There are rules about how it is to be determined, announced, challenged, altered. The agenda determines the meeting's scope and authority, and it introduces the governing principle of relevance. It puts some things beyond the bounds of discussion.

Within the framework of the agenda the chair regulates the flow of discussion—who may speak, in what order, for how long, to what question. The familiar parliamentary rules ("Point of order!") are not, from the perspective of the chair, devices by which partisans may balk the mind and will, but rather the means by which judgment can be better focused, expressed, and clarified. And, to this end, the chair may also protect the meeting against the bad manners of participants or the distracting intrusions of the audience. Free speech at a public meeting is a process shaped by limitations, presided over by a chair, advised by a parliamentarian, and supported by the sergeant-at-arms. Authority, law, and, if needed, the army. Our paradigm of free speech is a highly structured artifact. With this reminder to temper the feeling that, as we move into the consideration of forum government we are getting farther and farther away from good, old-fashioned town-meeting free speech, I turn to consider the general regulation of the forum.

The First Amendment may speak simply of "speech" whose freedom is not to be abridged; but we seem forced, in administering the forum, to develop distinctions between *kinds* of speech, kinds or classes of topics, and to regulate different kinds differently. All speech, simply because it is speech, is not treated alike.

Consider the rough classification: religious, political, commercial. Religious speech may enjoy special immunities and special restrictions. The truth of religious claims is not to be made a matter of official judgment, for example, and punishment for fraud may, consequently, be impossible. Proselytizing may escape some of the burdens of commercial selling. On the other hand, what looks like an ordinary meeting engaged in quiet speech might, if it is a religious meeting, not be permitted in schools or other public buildings. Political discussion? Perhaps, under special rules. Similarly, political communication and commercial communication are given different scope and are subject to different regulations. The borderlines between religious, political, and commercial speech may be troublesome, but the distinctions enter significantly into the governing of speech. Sometimes we may wish to distinguish, also, between public and private speech in terms of topic, and to govern these differently. Or we may establish a class of military, diplomatic, or national-security topics, topics that enjoy special treatment and which enter the forum, if they do, under heavy restriction. And, finally, let us recall that in many societies there is a class of forbidden topics, of illegitimate questions, of matters considered as beyond discussion—or general public discussion—as heretical, subversive, taboo, or merely private. The exclusion of such topics from the forum or their severe restriction to special contexts is sometimes an administrative charge, sometimes a matter of custom backed by unofficial pressure. The classification of topics, the discernment of kinds of speech, is an unavoidable part of forum theory. That "there is a time and a place for everything" recognizes the differences between the kinds of things we bring into discussion.

I now propose to take some slight liberty with "time, place, and manner." This familiar phrase is often used to indicate that some regulation of speech may be necessary. Even the staunchest "absolutist" defense of freedom of speech is likely to grant that regulation of time, place, and manner is not necessarily incompatible with free speech. So I borrow the phrase, although I may twist it a bit—harmlessly—to my own purposes.

I will not linger over a range of timing questions that are routinely familiar and, on occasion, even controversial—convenience in scheduling, respect for quiet hours, allocation of prime time, first-come first-served for meeting or parade permits—the timing, so to speak, of forum traffic. I turn rather to the question of unusual or abnormal "temporary" suspension of forum.

It may be that we give too little theoretical attention to the distinction between normal and abnormal times. Thus, we develop a set of rules and practices for the normal situation and expect that they should be able to survive some stress. But sometimes we have emergencies or upheavals or unexpected strains, and the question is whether we can continue to act as we would under normal circumstances or whether we can shift to the habits of abnormal times. Consider the theory of obedience. The normal theory is that, given a legitimate government acting with due process within the range of its authority, a citizen should obey the law. Then we add some strains: authority is usurped, or becomes tyrannical, or breaks down, or defies conscience; and the question becomes, do we continue to behave as if things were normal or do we shift to crisis, or state of nature, or "revolutionary situation" morality? The theoretical strain is over the differences between the attempt to project and insist on a single set of rules over the entire range of situations and the more routine acceptance of the fact of "emergency" that may call for the suspension of the normal rules. The Romans simply appointed a dictator when they thought an emergency warranted it. We can, sometimes, declare "martial law." But that is rare and extreme. Are there, we may ask, degrees of emergency permitting degrees of departure from the normal, or is it all or nothing—normal rules or martial law—, or is it, as some would have it, normal rules though the heavens fall? It is sometimes treated as a scandalous betrayal of freedom, virtue, and democracy, if the forum is ever suspended under emergency plea or pretext. I would not accept a pretext, but I suggest that we at least entertain the plea.

Apart from the existence of overwhelming disorder, there are two justifications offered for the suspension or adjournment of forum. First, to protect the integrity of the deliberative forum process itself. If the process of discussion is threatened by external pressures, or even by internal factional pressure, the suspension of forum may be the only way in which the community may be protected against the coercion of its tribunals. We have, I think, been altogether too tolerant of the coercion of meetings by mobs making "demands." It is probably better to dissolve the forum than to continue under threats. At least, dissolution until another time should be treated as an active alternative to continued operation in captivity or under undue pressure. (Poor Pharaoh kept insisting that liberation decisions he was plagued into making were not binding.) Temporary protective suspension of forum is sometimes justified.

Second, suspension of forum may sometimes be justified by the need to protect the community against consequences of its continued operation. The argument is that the community is too unstable, or immature, or too excited to enjoy without damage the verbal stimulation of the forum. In its most drastic form it is expressed as a theory of social tutelage according to which the community is "not yet ready" for a free forum and needs to be further schooled—a familiar postrevolutionary position. But even a mature community, habituated to a free forum life, may find itself in an irrational mood. A particular speaker on tour may leave a trail of riots;

a tense situation may need only a verbal spark to set it ablaze.
It may seem quite reasonable to suspend forum for a cooling-
off period, to cancel the speech, to ban public assembly, to
impose curfew. It would be foolishly doctrinaire to deny that
such situations, far short of justifying martial law, might
arise, just as it would be foolish, or worse, to resort too easily
and eagerly to suspension.

There are several familiar principles which bear on this
problem. The rule of "clear and present danger" limits suspension of forum to situations in which there is an immediate connection between the verbal activity and the legitimately preventable consequences; a general anxiety about
the tendency of speech to produce bad consequences is not
enough. And, complementing that, the belief that the forum
is self-correcting; that, if allowed to operate, it will correct its
own bad advice; that speech should be fought with speech.
This, as John Stuart Mill reminds us, presupposes a fairly
high level of rationality and maturity and some faith that the
teaching power has done its work.

Even on this assumption, however, the distinction between
normal and abnormal, between usual and extraordinary, between routine and emergency does not disappear. And the administration of the forum will require, beyond the temporal
considerations of forum traffic, provision for the suspension
of the forum for a time.

Turning from "time" to "place" we begin by noting that
here, too, there are many familiar issues over the place of
communication. Where may we not picket or assemble, or
march, or distribute, or accost, or display, or importune, or
preach, or demonstrate? Communication is not in place everywhere and what may be permitted in one place need not be
permitted in another.

But the forum has intrusive tendencies. There is always
someone with an urgent message about private or public or

divine business who sees everyman, everywhere, as a potential receiver of the word. Left uncontrolled, what scene would be without billboards, what park without its soapbox, what doorbell without its pushing solicitor? "Why not?" is the insistent demand, "Why can't I have free speech *here?*" The initial answer might be that there are plenty of other places, adequate to all our needs, more than enough, and that this is too much. But the appeal to other places as enough, even though there may be to any reasonable mind a more than adequate forum, is not likely to convince anyone with a particular urgency, and it has a rather feeble, unprincipled ring.

The attempt to put or to keep the forum in its proper place gains its significant support from two distinct, although sometimes related, principles: that a captive audience is not always fair game, and that places may be dedicated to special uses.

To some extent freedom of communication implies that "receiving" must have some tinge, at least, of voluntariness. We need not go to the meeting or turn on the set or read the paper. But we find ourselves in places of work, in parks or recreation areas, on buses and airplanes, in churches, hospitals, army posts, prisons, schools. And because we are, or must be, there for reasons related to what they are, we are, if we are to be made an audience, a captive audience. Must we be? Can we protect a person going about his business or pleasure from the message unavoidably imposed upon him without his leave or against his will? We cannot, obviously, require everyone's consent to everything that might solicit or force itself upon his attention as he moves about in public. But we are not altogether helpless. We can arrange for park areas without speeches, for scenic drives without signs, for the turning off of some junk mail, etc. And more significantly, we can provide that a group necessarily and involuntarily assembled for one purpose not be seized or used for another. As a professor, for example, I may have to attend a faculty meeting, called to discuss faculty business. But I and my colleagues have some right not to be subjected there, as a captive audience, to someone's views on anything not on the agenda.

Consideration for the rights of a captive audience may not seem to check significantly the exuberant expansiveness of the forum, but it is an argument for diverting forum activity to another place.

The second principle is, I think, more interesting and more important. There are some places that are dedicated primarily to use as a forum. There are other places whose primary use is something else and from which forum activity may be either altogether excluded or severely limited by the need to accommodate to the primary purpose. Parks, prisons, army posts, government laboratories, streets, and public schools have primary purposes other than, say, providing places for political discussion. Such discussion may be either totally excluded or required, if permitted at all, to conduct itself so as not to interfere unduly with the primary use. The latter alternative seems hardly disputable. After all, we build streets so that traffic may move, not so that it may be blocked by demonstrations during rush hours. If we permit demonstrations and marches, they must be scheduled and conducted with some respect for traffic. Similarly for other places. The forum is not licensed to destroy or seriously frustrate the army, the prison, the hospital, the laboratory, the school—on their own grounds; it must be required to be accommodating, at least, and to show that its activity is compatible with the other purposes at stake—prison or military discipline, serenity and quietness, concentrated work, etc.

Let us consider the case of the school more closely, since recent Supreme Court treatment shows, I believe, serious misconceptions. The Tinker case (*Tinker* v. *Des Moines,* 393 US 505) involved the right of schoolchildren to demonstrate their opposition to the Vietnam war by wearing black arm bands to school. "First Amendment rights," the Court said, "applied in the light of the special characteristics of the school environment, are available to teachers and students. It can hardly be argued that either students or teachers shed their constitutional rights to freedom of speech or expression at the school house gate." This has an engaging ring, but we either do shed our rights at the gate (schoolhouse or prison or office) or those rights are violently affected by the passage. Outside the school the student may read what he pleases or not read at all; inside, he must read what he is assigned. Outside the school the teacher has a right to campaign for his candidate; inside, he has no such right. Outside they may pray together; inside they may not. It is indeed, ridiculous to assert that we may exercise the same rights within as without the gate. If the First Amendment covers America in some sense, it must be understood to mean different things in different contexts, to permit different things in different places. As a teacher, I leave my right to agitate for my political views outside the classroom that has been provided for another purpose. And I consider it odd to suppose that students retain those rights there.

But I hesitate to limit the argument—although it goes quite far and perhaps far enough—to the need for the forum right to accommodate itself to the "special characteristics" or primary purpose of the context. So that, for example, I may indeed engage in political agitation in my class unless it can be shown to interfere with the purpose of the school. Or that I carry my right to make speeches into the park with me, unless it can be shown to interfere with someone's pursuit of another goal, that, perhaps, is to be at the hazard of someone's "balancing" of values. Can we not say simply that there are some places, even under the American flag, from which the forum is excluded? If we are not misled by pleasant images of rights, like Mary's little lamb, following us wherever we go,

this might be obvious. Can a community not say, "We will
tax ourselves to provide a park for recreation only and *no one*
can make a speech there?" Can it not build a school and
declare that while it is in session it is *not* to be used by anyone as a political forum? Does the schoolhouse gate signify
nothing?[36]

The issue can be stated as a conflict between the conceptions of an ubiquitous forum and of an adequate forum. I
have already suggested that a community must have a forum
"adequate" to its needs—however difficult it may be to translate that into particular terms. It is, as are so many things, a
matter of generous reasonableness. But sometimes the forum
is seen not merely as one great institution among others, but
as uniquely and fundamentally pervasive. It is as if the entire
country were paved with forum, supporting communication
everywhere, except, grudgingly, where it is overlaid by other
structures that cannot quite be forced to provide an open
easement to the universal verbal foundation. Less fancifully,
the defenders of the principle of the ubiquitous forum hold
that there is at least a strong presumption of the right to
forum everywhere, to be overcome only by a showing that it
would endanger (perhaps clearly and presently) some other
necessary or important social function. On this view, forum
is never, in principle, out of place. Neither view, however, re-
lieves government of the necessity of making regulations and
decisions about communication in the context of "place."

Time, place, and now "manner." In a simpler day, there
was oral face-to-face communication and the mass medium
was the printed word. Today we live in a sea of communication, buffeted by newspapers, magazines, books, Xerox, film,
tape, screen, tube, transistor, demonstration, vigil, sit-in, signs,
bill-boards, bumper stickers, sky writing, microphones, amplifiers, and bugs. The world can be admitted to the private
meeting. The transient remark can be instantly and endlessly
replayed, spliced into a variety of contexts. Silence is a boon

to the deafened mind, and thinking, more than ever, an effort to ignore.

The interests of privacy and of common deliberative reason have required a new range of support and protection in order to live with the new modes of communication—the regulation of electronic journalism and other forms of electronic intrusion and pollution. I will not linger over these problems that are accumulating into a special field of study, but will touch lightly on the old-fashioned question of "manners."

The struggle for attention, partly related to the new technology, has given rise to an array of imaginative devices. Sit-ins and similar demonstrations are treated as forms of communication claiming First Amendment protection. They tell us something, often—perhaps especially—when we do not want to listen. Even when "non-violent" they may be intrusive, disorderly, and ritually outrageous. And sometimes even violence—kidnapping, bombing, hijacking, "executions"—claims to be a necessary, and therefore presumably justified, way of telling something to an audience whose attention is hard to capture. A revolutionary mode of public discourse, a manner of speaking to a corrupt society, the only language it understands.

Opposed to all this there was, of course, a reactionary and sinister tradition of mannerliness in debate. Weapons were to be parked outside, and that scoundrel was to be referred to as that honorable gentleman. We were even expected to refrain from providing the background of boos, jeers, groans, or cheers, against which a speaker's remarks could be properly interpreted by those otherwise likely to be misled. "More light, less heat" was the motto. Temperance in utterance, it was said, is a mark of mastery; politeness, a recognition of the humanity of a temporary adversary. But all that was before we realized that politeness was hypocrisy and before we learned to recognize in intemperance the new sincerity.

It is probably quixotic to suggest that the forum be administered with an eye to manners. As the ancient motto puts it, "If you have nothing to say, it is necessary to shout." There is a forlorn hope that the school might inculcate the manners appropriate for human discourse; but it is probably necessary to reconcile ourselves to the fact that the participant in the public forum must acquire the arts of actor and animal tamer and that, inevitably, Pericles gives way to Cleon. Not even the university, that center of reason, can guarantee to a guest speaker on a controversial issue anything less than an ordeal; it cannot promise a fair hearing. I do not know what government can do about all this. It is not a trivial matter. Bad forum manners coarsen the public mind and subvert the human community. But how can we presume to chide the ardent heart?

Government is involved with the forum, I have been arguing, first, as sustainer of the constitutional structure of protected opportunities and, second, as regulator of "time, place, and manner." I will now try to show that it has a role beyond all this in aiding and protecting public and private judgment; that it enters into the life of the forum and is not merely the keeper of the ring.

Government is a primary producer of information and we expect it to provide us with much that we need to know. It conducts the census and publishes a broad range of statistical data and indices. Its agencies do a great deal of research and produce a flood of reports or studies. There are records of the work of courts, administrative agencies, legislative bodies, commissions. It also is involved in dissemination. Apart from its own publications it provides significant subsidy for the distribution of newspapers, magazines, books, and advertising through the postal service. The importance of the work of government as provider of information is rather astonishing.

I doubt that this is generally appreciated. The popular phrase is "government hand-out" and it suggests self-serving and unreliable propaganda. There is some, but the most cynical American would be forced, if he looked into the situation, to acknowledge that the derogatory phrase misses the point widely. Another popular perception seems to be that government is engaged in withholding rather than supplying information. It does withhold, but rather marginally and often for good reasons. The disparity between perception and reality may be illustrated by the case of the public library. In many communities it is the chief source of books, periodicals, research materials. It is supported by public funds and administered by public officials. There is an occasional scandal when a controversial publication is, under pressure, removed from circulation and the cry is "censorship" and "government keep out!" Keep out of what? Books and magazines do not grow in libraries, nor do they select and purchase themselves. Every item in the library is put there by a positive decision of government. Some bureaucrat decides to order this and not that. According to what policy? That is really an important question, but we seldom ask it. It is more exciting to think of government as the remover of books, not the supplier; and, thinking of government in that role, to demand that it cease to meddle with the mind. Neveretheless, government supports the life of the mind with a flood of information.[37]

In addition to supplying information we expect government to come to our aid by guaranteeing the truth of statements and by punishing misrepresentation. We expect that if a package says "12 net ounces" or "no preservatives added" the statement can be relied on, and we count on government to monitor the situation. The consumer movement is putting its weight behind increasing the scope of such government guarantees, and for fuller disclosure, for "the whole truth," and even for the truth in an unambiguous form—as in "truth in lending" laws and in "price per unit" rules. The mind, we think, has burdens enough and we welcome—even insist upon—this form of governmental support.

This, doubtless, is not what is meant when it is said that government has no business certifying the truth of statements. But we should begin there, because the problem of drawing a line between the area in which it may and the area in which it may not is not altogether an easy one. And *that* is the shape of the problem: *where* may government insist on truth and punish falsehood? We are forced to develop some distinctions. Religious statements? No. Commercial statements? Yes. Political? Official? Sworn? However we decide some particular questions, truth, misrepresentation, deception in utterance are not altogether beyond the reach of government in the forum.

There is more. In an earlier day most communication was local and we had many clues to aid our judgment in evaluating what was said. If you are familiar with a speaker, you may know his special interests, his character, his habits of accuracy or exaggeration, and make allowances. In an age of mass media and the professional thespian-communicator it may be necessary to restore some of the missing cognitive clues. Thus, government regulations may require that paid political announcements or commercials be clearly identified as such. The professional lobbyist must reveal his employer, the special interest of the advocate identified. Presumably, it helps us judge properly, to discount bias, if we know that the minimizer of smoking risks works for a tobacco company; so we should be told.

This is not altogether uncontroversial. Such interpretive aid—"consider the source"—may be said to divert attention from the truth of what is said, inviting us to commit a form of genetic fallacy. The truth of a statement and the motives of the speaker are two different things and we should be willing to accept the truth from any source. As the Greeks knew, Athena assumes many guises. But it is also the case that life is short; that we cannot prove everything; and that in practice we may and do consider the reliability or character of the source. It is at least a legitimate short-cut, and sometimes it is all we have to go on.

How far and in what directions government should go in unmasking special interest or bias has occasioned bitter dispute and created some reluctant bedfellows. The most energetic advocates of "exposure" have been, on the one hand, congressional investigators in the tradition of Senator Joseph McCarthy and the House Un-American Activities Committee and, on the other hand, the purity-in-politics conflict-of-interest crusaders. "You may speak," says the one, "but first tell us the political organizations you belong to." "You may speak," says the other, "but first list all your stocks and directorships." The indignant exposer of the establishment connection of a defender of atomic energy may object bitterly to the exposure of the antiestablishment connections of the advocate of unilateral disarmament. The distinction between political and economic interest and associations may bear on this problem, but for other reasons than relevance. The argument for exposing economic associations and not political associations is that freedom of political association needs the protection of privacy. That is a significant point. I am not sure that loss of economic privacy might not also deter political activity. However we decide this question, we should not pretend that knowing whether a person is a fellow-traveler or a communist or a fascist does not tell us something about his biases at least as revealingly as knowing his economic holdings. At any rate, government, to aid judgment, is asked to restore some interpretive clues to the situation by the exposure of bias—sometimes political, sometimes economic.

But bias exposure is not the only problem of interpretation aid. We know that the television screen presents us with only an edited version of what happens in the world. The principles that govern the editing process are vital to a viewing world; and while editing is a jealously guarded prerogative, there is a fundamental public interest in its operation. Recently, a congressional committee was denied access to material that would have enabled it to study the process by which an hour-long "documentary" was distilled from masses of material in a way that evoked charges of misrepresentation and misinterpretation. But the principles of editing cannot remain a private mystery forever. In providing aids to our interpretation government may someday cast a curious eye on those who are doing the editing and interpreting for us.

Government is even called on to assume a duenna's role. We are under constant seductive pressure to indulge our addictions—tobacco, alcohol, drugs, autos—and we may wish to consider whether it is possible to fend off some solicitation either by restricting it or by coupling it with warnings. Is it necessarily the case if an activity is legal, that urging such activity must also be permitted? Can we restrain or control persuasion to do what is permitted? Smoking is legal, but we require that advertising and packages display a warning. Advertising alcohol is under some restrictions. The fear of many about decriminalizing marijuana is that it would unleash massive promotional campaigns. Would it be possible to prohibit or strictly control such advertising?

Consumer chaperones insist, these days, not that advertising be prohibited but that it be linked with more and more disclosure that might have a chilling effect on use if not on speech. In a consumer civilization, "caveat emptor" has become rather tattered, and the brave new breed of consumer protectors seems bent on forcing government to play big sister. Initially, only in the commercial world. But there are signs that this is spilling over into politics.

There are two devices available: prohibiting or restricting the urging, and requiring that it be linked with information or warnings. The first seems to fly more directly in the face of the freedom to speak, while the second only requires more disclosure, interfering merely with the right not to speak. Both are chaperoning devices and may be viewed with mixed feelings.

Finally, besides what it does by way of informing, guaranteeing, exposing, balancing, and chaperoning, government enters the forum as a voice or many voices within it. Public officials, legislators and administrators, politicians, incumbent and aspiring, explain, propose, urge, defend, and advocate. They are active participants in the life of discussion which shapes the public mind. This is so obvious that I mention it only for the sake of completeness and, if necessary, to remind us that the proper image for the relation of government to public opinion is not one in which government, the "servant of the people," stands passively waiting for orders from its master; but rather one which recognizes that government is to lead, to discern needs, to formulate, propose, persuade, to enter as an active partner into the deliberative life of the community, into the process by which we make up our minds. Government is not, at this point, to leave the mind alone.

The government of communication must come to terms with several notions that may seem oddly opposed to the "speaking out" that expresses the openness of the forum—secrecy and silence. I turn first to secrecy.

The closed meeting is a normal part of forum or tribunal life. The jury deliberates unobserved; legislative committees may meet in executive sessions; the Defense and State Departments work in confidential huddles; school boards may consider personnel questions in closed session. The Supreme

Court, it should be noted, deliberates routinely in closed meetings to which only the justices themselves are admitted.

This general practice, contrary to the principle of "open covenants openly arrived at," seems to be a mark of an imperfect world and, presumably, when the lion lies down with the lamb all meetings will be open. Open meeting or "sunshine" laws are now, in the wake of scandal, rather popular; but they cannot eliminate the closed meeting altogether and they do not really try to do that. In some areas, to do so would be absurd.

To the ever-present cynic, the reason for the secrecy of the closed meeting is to permit the establishment to consummate its normal betrayal of the people or, if not that, to cover merely unprincipled corruption. (The same cynic turned revolutionary has been heard to argue that since revolutionary regimes have nothing to hide, they don't *need* open meetings.) But apologists will parade more respectable reasons.

First, the closed or restricted meeting encourages a higher quality of deliberation. It permits greater candor and genuine freedom of expression. It permits one to be convinced by others, to change one's mind without seeming to observers to be spineless or indecisive, to think out loud and suggest foolish things, to everyone's benefit, without the risk of appearing stupid to the general world. And when compromise is necessary, it can be worked out. In short, real work can often be done better in a closed meeting.

Second, some deliberations and the decision to which they lead would be frustrated or defeated or made pointless if prematurely revealed. Diplomatic and military matters come to mind at once, but there are less dramatic examples. Economic decisions, or decisions that involve timing and steps in sequence, or matters that require tentative prior investigation. Some good things could not be done at all if every step had to be taken openly.

And third, the closed meeting is, on some matters, simply more humane. Discussion of the relative merits or short-comings of candidates for employment or promotion or discharge preserves the dignity of all concerned when such discussion is kept relatively confidential.

These arguments are applied to public or official agencies trying to avoid the public spotlight. Less is needed to justify denial of access to meetings or gatherings under private or quasiprivate auspices, even though matters of public concern may be considered there. There are gatherings for adults only, for men only, for women only, for members only, for fellow ethnics only, and even for invitees only—for which some justification may be given but for which freedom of association as determined by whim may be sufficient. We may be surprised, however, to find that some exclusive occasions that are thought of as private may not be so private after all and may be required to open up.

I begin this discussion of secrecy with the closed or restricted meeting rather than with the secret item, the secret bit of information, or the dark secret, because I think the restricted or confidential character of deliberation is more fundamental than "secret" as a quality of information. The confidential *process* protects dignity, but not because it conceals "undignified" facts. Secrets, when they are ultimately revealed, are usually quite banal—like an old newspaper—and get their point not, I repeat, from some striking quality, but from their significance in an ongoing process. This is why it is so easy to ridicule classified information when it is subsequently declassified, officially or unofficially. "Why it's just a bunch of weather reports for Normandy beaches! *They* know what their own weather is like!" For the most part, secrets are temporal; they have a lifetime and a rate of decay and it is expected, at least with government secrets, that they will be yielded up to history, move from the sealed file to the public archives when they have been disarmed by the passage of time. They may contribute, ultimately, to our enlightenment in contexts other than that in which they were generated and merited, or claimed to merit, concealment. The existence of secrets—the deliberate withholding of information from general knowledge and communication—is an inescapable aspect of the forum that, as it must, has a place for the closed meeting.

We should note that the problem of secrecy appears differently from two different perspectives. There is the secrecy that government preserves from the eye of the citizen; and there is the secrecy, which we prefer to call privacy, that the citizen preserves against government. The sovereign keeps things from the subject; the subject keeps things from the sovereign. On the familiar principle that the only justified secrets are my secrets we can anticipate the following: *Government:* "How can I serve you if you keep things from me and keep prying while I'm trying to work?" *Citizen:* "How dare you hide things from your master and who said you could pry into my private affairs?" There is a constant clash, in both directions, between the claim of a right to know and the claim of a right to conceal or withhold.

There is indeed, in some circumstances, a right to know, but it is not defined by a desire to know, by curiosity—idle or active. We cannot demand answers to our questions, as a matter of right, simply because we want answers. We are entitled to know by virtue of some functional status; the right to know is tied to the need to know. The pattern is simple in the case of a governmental assertion of a right to know. A government agency—a grand jury, a court, a legislative committee, an administrative agency—is given a function to perform. In discharging its responsibility it claims the information or testimony it needs, sometimes in the face of a strong reluctance to disclose. It may have the power to subpoena records and witnesses and to punish non-cooperation or non-disclosure. It is not a roving curiosity that supports the demand for information, but the relevance of what is sought to the legitimate public purpose at stake. The right of the citizen to know about the actions of government has the same basis. The argument is that since he must, as a member of the electorate, pass judgment upon politicians and policies, he needs to know, and therefore has a right to know, what is being done and how it is being done.

But the right to know, even when based on the claimed need to know, does not penetrate everywhere. I have already touched on the closed meeting in which the deliberative process is shielded from scrutiny and in which even the results may be kept secret for a time. The justification for the right to withhold, as for the right to know, is the necessity of the legitimate function.

In addition, there are a number of relationships that are impervious, in varying degree, to even a strong need to know—relationships of a privileged character. In order to protect the intimacy of the marriage relationship one may withhold testimony relevant to the guilt of one's spouse. A lawyer may not be required to testify about what his client has revealed to him, even though that would be relevant to the trial. Similar claims are made and, in varying degree respected, for the privileged character of the relation of priest to penitent, doctor to patient, teacher to pupil, and even of journalist to source. And, of course, we may grant to the individual the privilege of not being "a witness against himself," the so-called privilege against self-incrimination. Even a legitimate need to know does not necessarily or easily override the values embodied in such privileged relationships. The right to know may be overridden by the right to withhold, to conceal, to keep secret.

Confidentiality, the withholding from public view, may be suffering too much disrepute. Even where public business is involved, secrecy, as we call it then, may serve many values. Where it is non-governmental, confidentiality is seen as privacy, and we are disposed to defend it. It is not merely our immorality that makes us shudder at the thought that we are everywhere, always, being taped and filmed, broadcast and shown. We would mind, even if we were good.

Modern technology makes privacy and secrecy more difficult to protect, and it is unlikely that we will abandon technology. So, if we wish to preserve our sanity—public and private—we must learn to restrain any capacity to tune in on things. We must remember, in our lust to know, that curiosity establishes no broad right to knowledge and that the right to know, even based on need, can be too lightly claimed.[38] It does not override, in its hallowed cognitive arrogance, every other human value.

Seen from a slightly different perspective, questions of secrecy and the right to know reappear as the question of the right of silence. To what extent can an individual withdraw from participation in the life of communication, keeping his experience, his knowledge, his judgment, to himself? Must he speak if he does not want to, disclose what he does not want to reveal? Is he sovereign over his own mind, sharing it or not as he alone sees fit?

The individual is, of course, subjected to the teaching power willy-nilly, but compulsion beyond that may be only sporadic and informal. In some polities he may be required to record his judgment at the polls; in some, he may be required to participate in religious or ceremonial avowals or disavowals. He may, more generally, be required to contribute his knowledge, to tell what he knows, before a court or investigatory tribunal. There is, in short, no generally recognized option to refrain at one's simple pleasure from speaking. In some circumstances—witnessing a crime is the obvious example—there may be an enforceable duty to communicate.

Disclosure of one's state of mind may be, if not compulsory, a condition of access to opportunities. One may be required to submit to examination and cross-examination, to submit to and cooperate with clearance procedures if he is a candidate for a position of responsibility or trust. He cannot always say that what he thinks is his business only.

More interesting difficulties arise over the right of a possessor of knowledge, a scientist or technician or expert, to withhold his knowledge from particular persons or from the community as a whole unless the knowledge is put to uses of which he approves.

Is it not galling to be a shoemaker, to make a beautiful pair of shoes, and to see the client stride off heedlessly, in those shoes, on the road to ruin? Or for a doctor, to cure a patient, and to see him return to his evil ways? Or for an architect to waste a beautiful house on sordid lives? How much more—but why, really, more?—must it gall a scientist or specialist to see his work put to foolish or destructive or immoral uses. Why should he tell people how to do things if they are bent, in this case or that, on doing the wrong thing? Why not insist that enabling knowledge be used only for good? Thus is raised the question of the social responsibility of shoemakers, or longshoremen, or of the scientist.

But surely, what is meant is that the scientist should be aware also of his responsibilities as a member of society, that he should take part in public affairs, vote, speak out on public issues and especially on those public issues where his knowledge is relevant. It cannot mean—certainly not at this late date—that a scientist "knows better" about anything but his own specialty, that his superior knowledge warrants his enjoyment of a veto power on the use of his knowledge. Something still remains of the distinction between how or whether something can be done and whether or not it should be done. The specialist is not a sage. The military expert may know all about warfare; that does *not* include knowing what is worth fighting for. We put him, for better or worse, under civilian, political, control and we would be a bit startled if, in the name of the social responsibility of the military, the army would fight only when *it* thought best. The case of the scientist or intellectual is no different. A physicist or chemist or biologist or mathematician has no privileged views about morality or social policy, and if he thinks he has, he merely adds pride to ignorance. He is entitled to enter the political arena like anyone else. The selective withdrawal of service is another matter.

We must, however, place this problem in the context of the institutions within which a particular society provides for the availability of knowledge. While it is not the case everywhere, there is, with us, almost a free market in knowledge. That is, an expert may contract to provide his services to private employers, to a variety of government agencies, to movements or causes; and he can move from one to another at his own discretion in response to market conditions. He can go into business for himself or sell himself as a freelance. In this way, choosing, changing and even quitting, he can in good part accommodate the use of his knowledge to purposes that he finds good or tolerable and even engage in the partisan practice of his art.

In the case of the public functionary the demands of the office may involve more strains. He may be required to serve all comers. Or shall we say that a county agricultural agent may refuse to give requested advice because that would help a hippy commune maintain a way of life he regards as an abomination? Or that the army doctor may refuse to teach about first aid for burns because he thinks it's a bad war? Or that a Catholic doctor in a public hospital may, at his discretion, decline to answer questions about birth control? Surely there must be rules about this sort of thing. The cognitive functionary cannot be permitted to discriminate according to his private moral schedule. He must advise the hip and the square, the ecologist and the builder, the hawk and the dove, the Catholic and the Protestant. In an obvious sense, what they do with his advice is none of his business.

Where this is intolerable, resignation may be accepted as the device for withdrawing knowledge-service when the expert strongly disapproves and cannot veto the use to which it is put. But resignation is a severe option and deprives one of the chance to do some good. It may entail complete separation from a hard-won art and way of life. Often it is not really an option at all, and one must stay and act contrary to his convictions (or perhaps become a devious saboteur). But justice may not permit even his well-intentioned discrimination.

Sometimes all options may disappear and the services of the mind may be drafted. There is, so far as I can see, no good reason to say that while a person may be required to risk his life in the military service, he cannot be required to serve by contributing his special knowledge. The community may need his knowledge more than it needs his rifle. And it may be entitled to it.

The case for the social claim to the services of the mind is based not only on the social need for knowledge but also on the social character of its generation and development. A scientist, for example, may have gotten his early education in public schools and been supported through graduate school by public funds. He works in government-supported laboratories with expensive facilities. He may live his life within the hospitable embrace of public institutions that have enabled him to pursue his bent and interests. It is rather absurd, under these circumstances, to act as if his knowledge, his science, is, in any significant sense of ownership, "his," to do with altogether as he pleases. It may be more appropriate to think of the scientist, the cognitive professional, as a deliberately cultivated type of social functionary—trustee, not owner—to whose special skills society has a valid claim even under protest. The claim of a right to withhold knowledge even when it is drafted draws no special validity from the fact that the mind is involved. It is, at most, another claim to refuse service as a conscientious objector, enjoying whatever respect a society may give to that claim in general.

There is, in short, no absolute right to remain silent. The individual may, in a variety of situations, have a duty to enter the forum, to communicate, to make his mind, his experience, his knowledge, available to the community. Allowable silence is a matter of policy.

I began the discussion of the forum with an emphasis on the aggressive character of speech, on forms of verbal assault. I had better not let it go without saying that speech is an instrument of communion and understanding, that it is the best hope of brotherhood and peace. The emergence of its polemic character is, I think, a defeat for its proper, its deeper, irenic mission. There is always, when speech is used as an instrument of warfare or conflict, a lurking sense that it is being degraded and misused. Language, like the cattle of Helios, is dedicated to the gods, and if we divert it to our lesser purposes we damn ourselves to destruction. Just as misunderstanding is a failure of intended understanding, so the aggressive mode of speech is a failure in communication.

It is instructive to consider the condition of the forum from this point of view and to attempt to locate it on the scale of degeneration. Near the top of the scale—and it does not matter whether we think of it as in a past golden age or in a golden future or as a timeless form—is the great deliberative assembly in which virtue and wisdom display themselves calmly in the dignified service of the common good. This is the assembly at which the invading barbarian gapes with little understanding and a touch, perhaps, of superstitious awe before he sends it packing.

When disinterested deliberation about the common good breaks down, it is supplanted, dominantly, by the bargaining transaction. This itself, in its higher form, has a genuinely civil tone. It understands mutuality—that a good bargain is good for both parties. It accepts interdependence with good humor and recognizes that in a culture that lives by the bargain, we are all in the same boat. It knows the hard bargain but it acknowledges a bargaining morality. It is a possible mode of decency.

This, too, is a forum that another kind of barbarian invades. Peering with half a wit he discovers interdependence and overlooks mutuality. The mutual—essentially good-humored—bargain turns, as in a nightmare, into an extortioner's demand. The smile becomes a wolfish grin. Interdependent, right? You can't get along without me? So here are my demands and if you refuse, I'll drown you in garbage, or let your children roam the streets, or let your fires burn, your goods rot on the pier, your red-tape tangle. We discover "The offer you cannot refuse." Communication is a statement of non-negotiable demands, a snarl.

How close this is to violence, piracy, and terror is seen when we face, as the inevitable next step, the communiqué from the enemy who has seized hostages or planted bombs and tells us what will happen "unless" . . . "It's just pressure, isn't it?" say the terrorist apologists. "It's the only language you understand. And how does it differ from what corporations and unions do? Only in degree, not in kind."

So, by a succession of subtle parodies, sometimes by indiscernible degree, the forum swings from reason to violence and perhaps, if we are lucky, back again. There may be something to the view that, violence for violence, the violence of the forum is a benign surrogate for that of other arenas; that it is, with all its confrontations, a "moral" equivalent of war. No doubt. But surely we must study the process of degeneration and push in the other direction when we can. The governing of the forum should not be indifferent about the decay of reason or unconcerned about the effects of its institutional arrangements on the mind.[39] The hope is that the forum can become the sustainer of a mode of communication that is more than a form of conflict.

The range and complexity of problems facing a society that seeks to provide a life of communication adequate to its tasks do not permit the luxury of treating freedom of speech as simply a native right of self-expression limited only by clear and present danger. The freedom of speech is like the freedom of the city. To be given the freedom of the city is not to be permitted to do whatever one wishes, but is to be granted the status of being under the city's law. To be granted the freedom of speech—and it *is* a kind of bestowal—is to be given a place in a rule-governed forum, the institution within which and under whose protection the citizen may share in the reflective and deliberative process by which the community seeks to govern itself.

Postscript

In each of the areas in which government is involved with the mind we should look first to the basic social institutions—the forum, the school, the laboratory, etc.—and consider whether, or the degree to which, government is necessarily involved in constituting, supporting, and giving some shape and direction to those structures that are so necessary to the very existence and well-being of any society. There is, I believe, a degree of supportive involvement with the mind-related institutions in any society, regardless of the policy orientations that reflect more directly the peculiar values and commitments of a particular social order.

We are led then to consider the deeply revealing issues of a constitutional order—public ownership, monopoly, control, delegation or diffusion of authority, pluralism—that present us with stark contrasts between open and closed, democratic and dictatorial, free and totalitarian societies.

And, finally, beyond the problems at the basic structural or constitutional level we encounter, even in democratic societies, particular questions of policy: shall certain kinds of research or teaching or modes of expression and communication be permitted or required, for example.

At every level there are questions of legitimacy and questions of wisdom. Acts that are legitimate may, nevertheless, be quite foolish; and at times the requirements of legitimacy may prevent us from taking the wiser course. Also, as we know, what is illegitimate in one system may be quite legitimate in another. Thus, an established church may be illegitimate under the United States Constitution; it may be quite legitimate elsewhere, as in England. It is a familiar mistake to confuse one's parochial rules with the higher law; but that is not to say that there are no universally applicable principles of higher law. My caution is simply that we take a more modest view of the ultimate moral claims of any particular (one's own, especially) constitutional structure, that we recognize the distinction between universal and local legitimacy, and that we remember that wisdom is another matter still.

It was not my intention to dwell with special emphasis on the American context or to develop a position uniquely appropriate to our constitutional system. But that is what I am most familiar with and I have not resisted the tendency to draw on American examples or to put the argument in American terms or to think of myself as addressing an American audience, even though I consider my general thesis as universal in scope. We are, among the countries I can think of, the most hospitable to the dogma that government has no business meddling with the mind. And while our actual involvement is, as I think I have shown, very extensive, we are at one extreme of the spectrum in our rejection of direct government control or monopoly. It is not difficult to find countries in which my thesis would be taken as so obvious as to evoke only puzzlement over the fact that anyone should bother to argue or even assert it, and in which government monopoly and direct control over school, forum, and laboratory would be the normal arrangement. I must seem, from that point of view, provincially American. But if it is not amply clear, let me make it so. I do not yearn for simplicity. I do not favor government ownership of or monopoly over the institutions of the mind. I am a child of the American system and I enjoy it, respect it, and marvel at it. It has its sins, but it did not grow in a day, and I do not sit, like Jonah, in the shades of academe, awaiting its promised destruction.

The special structure of the American system is a product of our history and culture and, of course, it is still evolving. But any treatment of the subject must make some bow in the direction of the First Amendment whose text and gloss are so crucial to our understanding. And to some misunderstanding. How, then, does the First Amendment affect the thesis that government is legitimately concerned with the mind?

First, I must point out that the involvement of government in ways I have discussed has developed within our constitutional system and under the aegis of the First Amendment. It has not condemned the public school as unconstitutional nor struck down the F.C.C. nor nullified the principle of research grants, etc. It is compatible, on the whole, with what we are doing—or at least the Court, our authoritative interpreter, has said so in response to constant appeal. Much has been prohibited, but what we have is permitted. I do not wish to argue the merits of this or that decision. My point is that at least what we now do we do under the First Amendment. This may be nothing more than a reminder that a prohibition against the abridgment of the freedom of speech is not to be taken as a prohibition of aid, support, or even some regulation. The amendment, in short, has not put an area of concern "out of bounds" for government; it prohibits certain kinds of action.

Nevertheless, there is some tendency to accept as canonical judicial rhetoric about "the sphere of intellect and spirit which it is the purpose of the First Amendment to our Constitution to reserve from all official control." This does suggest an out-of-bounds domain and I can, perhaps, indicate the source of the misconception.

The First Amendment, when it was adopted, did indeed have a jurisdiction-denying character. It was a denial of the

authority of Congress, or the Federal Government, in the domain of religion, speech, press, assembly. Authority over these
things was to remain in the hands of the States. This was a
matter of deep concern and we can capture some of the quality of that concern if we consider a hypothetical contemporary example. Suppose that problems of world peace, justice,
and of economic life lead us to a reluctant willingness to consider a rewriting of the United Nations Charter, increasing
the governing power of the United Nations. Nations would
be naturally apprehensive about the possible effects upon
their own cherished cultural, religious, and political institutions and, concerned to protect their autonomy in these matters, insist upon something like a First Amendment declaring
that the new government, the General Assembly or the United
Nations "shall make no law" respecting them. This would indeed be a denial of authority in a domain. But not a denial of
authority to *all* government; only to a particular government.

The adoption of the First Amendment was, making due
allowances, essentially in that spirit. The standard histories
record that "One of the amendments which the Senate *re-
fused* to accept—the one which Madison declared to be the
most valuable of the whole list—reads as follows: 'The equal
rights of conscience, the freedom of speech or of the press,
and the right of trial by jury in criminal cases, shall not be
infringed *by any State*' . . ." (my emphasis added) and "In
spite of the deliberate rejection of Madison's proposal the
contention that the first Ten Amendments were applicable to
the States was repeatedly pressed upon the Supreme Court.
By a long series of decisions, beginning with the opinion of
Chief Justice Marshall in *Barron v. Baltimore* in 1833, the
argument was consistently rejected" (Constitution of U.S.A.
Revised and Annotated, 1952, p. 750). Constitutional scholar-
ship has not really shaken the view that, when it was adopted,
the First Amendment was a denial of authority only to the

Federal Government. The States retained authority "in the sphere of intellect and spirit."

It is also well known that the post-Civil War amendments, notably the Fourteenth, altered, in some respects, the federal system. It was not immediately clear, however, that the situation regarding the First Amendment was altered. It was not until well into the twentieth century that First Amendment limitations began to be imposed upon the States, and then only in a halting and puzzling manner. It was discovered, for example, that "liberty" in the due-process clause of the Fourteenth Amendment embraced a "liberty of speech" of which a person could not be deprived by a State without "due process of law." The subsequent judicial treatment of freedom of speech (or of the First Amendment generally) is among the more complicated developments in our judicial history. Its moving theme is the attempt to impose First Amendment limitations upon the States.

The discovery that liberty of speech is included in the due-process clause of the Fourteenth Amendment suggests a simple possibility: that the First would still be read as a "hands off" addressed to the Federal Government while the Fourteenth is considered a "due-process" admonition to the States. But that is too simple to be real. The struggle to bring the States up to "higher" standards has led to a popular acceptance of the view that, somehow, the Fourteenth Amendment "transmits" and makes the First Amendment applicable to the States. On this view the States cannot do, in this area, what the Federal Government cannot do. And it is only a short step from the "Federal Government must keep its hands off the mind" to "all government is excluded from the sphere of the intellect and spirit."

I have no special quarrel with our use of exegetical resources in the struggle to sensitize our administration of the life of the mind. Nor do I propose a quixotic return to a fundamentalist interpretation of legal concepts. I am concerned, however, with the intellectual damage that can be done in the process of seizing whatever is at hand to fight our ad hoc battles with.

The upshot of the "transmittal" theory of the First and Fourteenth Amendments is a virtual supplanting of a two-level by a single level theory—both Federal and State governments more or less subject to the same constraints. But it must be remembered that there are two statements governing the situation: "Government (substituting that for 'Congress') shall make no law . . ." and government shall not deprive . . . except "by due process." The same merging which inches the States toward "shall not" also moves government away from the "Absolute" *shall not* toward a mere due process—modified by a shift in presumption. Or, condensing the story, we are left with "government shall not act unnecessarily or unreasonably in the area of speech." This is, I think, if we look around us, about where we end up. And it is where we must end up if the alternative is that "government is excluded from the sphere of intellect and spirit."

This may come as a shock to the naïve and even fuel the disillusionment that seems to accompany the revelations of casuistry. Where have all the absolutes gone? Education is burdened, or perhaps only challenged, by this situation. But the result is that when the First Amendment is properly understood it does not stand as a theoretical obstacle to the view that government is legitimately concerned with the mind. We are badly in need of fresh studies, from this point of view, of conventional civil liberties doctrine.

It is not easy to judge the success with which government discharges its mind-fostering function. There is no simple unchallengeable pattern of sanity or mental health or wisdom against which we can measure ourselves or others. But we are aware of "failures"—in education, in communication, in awareness—and this sense of failure is rooted in the assumption of a "rightness" toward which we move, however gropingly. The normative study of the life of the mind must provide the theoretical underpinning for the government of mind, and this, perhaps, is what Plato hinted at long ago in the conception of the philosopher-king.

For a democracy, there is a leading clue. It is not enough, although it is necessary, that sanity is pervasive. The public mind must be seen as the mind of the Sovereign and cultivated into fitness for the dignity of that office.

Notes

1. *John Stuart Mill's protective principle*

Mill's principle, invoked as a protection against the "tyranny of the majority" in a democracy, as well as against government generally, differs from the principle under consideration here:

> . . . the sole end [he says] for which mankind are warranted, individually or collectively, in interfering with the liberty of action of any of their number, is self-protection . . . the only purpose for which power can be rightfully exercised over any member of a civilized community, against his will, is to prevent harm to others. His own good, either physical or moral, is not sufficient warrant . . . The only part of the conduct of anyone, for which he is amenable to society, is that which concerns others. In the part which merely concerns himself, his independence is, of right, absolute. Over himself, over his body and mind, the individual is sovereign.

Mill develops his famous argument in interesting and powerful ways and *On Liberty* should be read and re-read. He is so deeply concerned with liberty of thought and expression that I find myself saying, "even John Stuart Mill . . ." as if that, surely, puts the matter beyond further rational dispute.

Note, however, that in drawing the line at "what concerns others" Mill asserts the individual's residual sovereignty over both "body and mind." "Each is the proper guardian of his own health, whether bodily, or mental and spiritual." And note also his assertion that "this doctrine is meant to apply only to human beings in the maturity of their faculties. We are not speaking of children, or of young persons below the age that law may fix as that of manhood or womanhood . . . [or of] those backward states of society in which the race itself may be considered as in its nonage." (*On Liberty,* Chap. I)

That government has no authority in the realm of mind is different from the principle that government has no authority to interfere with an adult "for his own good." They are both attempts to state in simple form a principle by which the authority of government is to be limited. "Hands off the mind" has no patron saint, but it is, I believe, a familiar part of popular consciousness.

2.   . . . *the imagery of compartments* . . . .

We are deeply indebted, I believe, to religion, to churches, for their role in seeking to impose limits upon secular authority. But while religious claims may be asserted for the sake of or in the name of the spirit, they are not really claims to freedom "in spirit" only. Of course, doctrinal autonomy is normally claimed. But doctrinal freedom and the freedom simply to believe are not the limits of the religious claim. Freedom of conscience is the freedom to honor conscience by appropriate action. It is the secular power that is likely to say, "Believe what you please, but do as I say"—offering freedom of thought in exchange for authority over action. Religions, quite sensibly, are seldom content with this bargain and insist on a piece of the action. They reject relegation to a spiritual realm beyond relevance to overt action.

I like the "two-city" notion—heavenly and earthly, divine and human, invisible and visible, spiritual and carnal—but the point, of course, is not to see them as in different places but as in the same place. Not as if one were Los Angeles and the other San Francisco so that we can deny to the mayor of one any authority over the other; but as something more like a coterminous city and county in which the same thing can be seen now as city and now as county. The trick in all this, if we are to preserve the sense of duality, is to keep it from degenerating into territorial distinctness marked by walls of separation. The mistake is to suppose that we can put a fence around a spiritual area within which the soul can go its own way, leaving the governing of the body to the ruler of the other area. On some such misconception as this it might be said that government is not to intrude into "the realm of intellect or spirit."

3. . . . *the sense in which the mind is private* . . .

I have no quarrel with "inner" or "private" as ordinarily applied to the life of feeling, thought, reflection; only with the implication that "private" in this sense means "beyond the proper concern of *public* authority."

Privacy enters the picture because of the apparent unsharability of some aspects of awareness. Anyone who is permitted to look can see my broken tooth; but the pain, if there is any, is mine alone. Something can be publicly displayed. Something cannot be; it is unsharable, *mine alone.* This peculiar private status was fused over some years back when it was being argued that public verifiability was a condition of meaning and that, therefore, "he is in pain" should either be taken as meaning that he writhes, quivers, howls, displays cuts, bruises, etc.; or, if not something like that, must be considered meaningless, since not publicly verifiable. Only the publicly verifiable was thought fit for public notice. Dogged reactionaries, as I recall, insisted on the reality of feelings of pain and pleasure (more than the purr!) and the rest of the furniture of consciousness even though, when pressed, they acknowledged that such awareness was private.

But if consciousness—the toothache, resisting reduction to something else—is "mine alone," it does not follow that it "concerns me only," is of no public concern. That is another question. If we believe, with Mill or Dewey or almost anyone, that public authority is legitimately concerned with what is broadly consequential, then the question is whether beliefs, ideas, feelings, are in fact consequential and thus not, in the sense of "concerns me only," private and beyond the scope of public authority.

I do not stop to argue this question here—if it needs argument. At this point I only note the contribution to confusion over the relation of government to the mind made by a tendency to slide from "mine alone" to "concerns me only."

4. *Government acts in a variety of modes . . .*

Why do we persist in thinking of government in terms of images that make its involvement with the mind seem inappropriate, if not horrifying? There is the tradition, exemplified by Thrasymachus, stripping government of the aura of authority and reducing any claims of rightness to mere force. Government as conqueror or master concerned only with its own aggrandizement. This may get transformed into the conception of government as the keeper of the peace, carrying with it, as I've suggested, the imagery and apparatus of the criminal law. But beyond that? The guardian, the trustee, the provider of welfare all seem steps in the direction of "big brother" and *that,* in its Orwellian sense, is stripped of all warmth and of any intimation of guide or mentor. On the whole, the attitude is grudging, suspicious, cynical, and this is said to be the only safe attitude.

Nevertheless, government is more than army, police, court, and law-maker. It is the friendly mailman, the fireman, the social worker. It heals in its hospitals, and, above all, I think, it is the public-school teacher. It is foolish to insist that government is *really* only a coercer and that everything else it seems to do is only a fraudulent attempt to disguise its essential nature. Even if education is compulsory, education is not compulsion—any more than compulsory medical treatment turns the art of medicine into a form of force. Our government is healer and teacher as well as policeman.

5. *. . . community is a condition of mind . . .*

A community exists only where there are shared understandings, a special condition of mind. Consider a fundamental institution like property, reflected in the conception of "mine." To learn the language is to get some sense as to how "mine" is used. To understand that is to understand something of what may or may not be done, or what I may have or keep or use. There is an apple; but whether it is mine depends on something like finder-keeper or grower-keeper or buyer-keeper or even, perhaps, needer-keeper. The "mineness" depends on some social understandings without which there may, perhaps, be an apple, but no belonging. "Belonging" is a fruit of understanding, convention, custom, agreement. Property is a state of mind. So also for the whole world of proprieties, rights, obligations, statuses, offices, agreements—habits and conditions of mind. Strip *that* away and nothing that is distinctively social or human is left.

Thus, "community" is a condition of mind. How strongly this needs to be asserted or defended depends upon how strongly the reductionist tendency is operating at this point—that is, the tendency to say of something that it is merely something else. We have been through many reductionist episodes: biology is really merely chemistry; psychology is really merely biology; morality is really merely psychology; the mind is really merely the brain, etc. Here it would be, "the community is really merely a complex pattern of observable animal behavior completely describable without use of mental categories" or, perhaps, "describable in terms of psychological categories that do not include such dubiously normative or moral notions as rights, obligations, agreements, commitments . . ."

Established disciplines, by virtue of having become established, have survived their reductionist crises although the normative treatment of the polity may still, in barbarian reaches of academia, feel apologetic in the face of rude remarks by "operationalists" who dimly trace their philosophical orientation to outmoded forms of positivism. It is probably pointless to take the argument seriously. It should be obvious, upon reflection, that without a special condition of mind there is no community, no political community, and the brash "it is something else" should be brushed aside with "nothing is merely something else." Man *is* a political animal, incompletely described except through the categories of the polis. When we grasp the implications of that, we understand what it means to say that community is a condition of mind.

6.   *dependence of undertakings upon understandings*

This can be presented in both a modest and a grandiose form. Modestly, it is suggested that doing something, carrying on, playing a part, acting and enacting, requires that we see the next step and the next after that as strung together by a sort of plot. It is the knowing what is called for, understanding what one is up to, knowing where one is headed, that makes sense out of *this* as the right thing to do, as the next step in a sequence, present to the mind, that runs from past to future. Without the thread, the plot, the pattern, the objective—all kept in mind—action loses its point.

More grandly, we speak of intellectual foundations and spiritual presuppositions. For example, Walter Lippmann's theme in *The Public Philosophy* is that parliamentary institutions can work only within the framework of a particular tradition of civility, a philosophy of Natural Law. That provides the proper parliamentary frame of mind, the necessary spiritual basis without which parliamentary institutions make no sense. Or we may be told that democracy makes no sense except on the presupposition that all men are equal or rational. Or that honesty rests on the belief that there is a God. My favorite example is from Dostoevsky: the old General overhearing in his club some fashionable expressions of atheism, stalks out exclaiming, with a beautiful series of mental leaps, "If there is no God, how can I be General?"

Well, there is something to all this. There is a parliamentary frame of mind without which parliamentary institutions make no sense; there is a democratic frame of mind. This is seen strikingly when there is an attempt to impose (or to borrow) parliamentary or democratic institutions where the spiritual groundwork has not been prepared. A parody, sometimes amusing, sometimes tragic.

The point of all this is that a community, a complex structure of enterprises, has a great stake in the condition of the context of understanding—the mind—that sustains it, that makes sense out of its life. It is odd to suppose that the right to "defend" itself does not include a rightful concern for the mind.

7. *The school board is, in principle, legitimate.*

It is not necessary that a community establish a public school system to cultivate minds. It may leave education to the family or to private corporations or to churches; it can neglect or ignore it. But if it decides to undertake public education under appropriate governmental control, if it establishes public schools, supports them with taxes, if all this becomes a normal part of the constitutional order whose legitimacy is accepted even as we struggle over particular questions of policy—what are we to say if someone announces that the political and the mental are distinct and that it is illegitimate for government to operate a school?

Why? Is education irrelevant to the community's purposes? May it not provide for its needs? Must it be dependent upon spiritual alms? What principle of legitimacy does the public school violate? One can imagine all sorts of advice: Government is best that governs least; don't do anything for anyone that he can do for himself; just protect life, liberty, and property; don't try to do good, just prevent harm. But these old saws do not define legitimacy.

Legitimacy is a difficult notion, and I will not attempt to define it. But I will offer, at least, a clear case. My university—a state university, established in the Constitution of the State of California, supported by tax funds, commissioned to do work in the "sphere of intellect and spirit"—is a legitimate institution. Any analysis of legitimacy that would produce the result that, because it is what it is, it is therefore illegitimate, would be as mistaken as a definition of "dog" that banished Lassie from the kennel.

8. *. . . government . . . (state, society)*

A word is probably in order about the ordinary use to which I put "society," "community," "state," and "government" in the course of discussing the authority of government over the mind. Authority, of course, is distinct from power; it is at least legitimate power; and I assume that that is understood.

Society and community may be considered as sociological concepts by which we refer to groups or persons in various institutional settings, related in various ways, without necessarily attending to or being concerned with the specifically "political." The state is a community or society considered in the light of its political structure; it is a society or community organized under government (or under *a* government). Thus North America is not a state; nor is Europe. We may speak of North American society or of the European community. North America (also a geographical expression) includes Canada, which is a state and the United States, also a state, as Europe includes the states of France, Italy, etc. We may, for many purposes, be unconcerned about Canada as a state and speak of Canadian society, or of Canadian family patterns, or Canadian linguistic communities. But we can also turn our attention to some "political" aspects and consider Canada as a state, organized under its government. When we do so we are not abolishing or doing violence to society and community; we are simply paying attention to something. State and society are not distinct or opposing entities anymore than I, as voter and as tennis player, am two distinct persons.

"State" and "government" are not the same. A state *has* a government; in addition, it has the governed, the subjects, the members. It is a bit difficult to conceive of a government with no one to govern. And it is strange to identify state and government. A state has government and governed. Thus, the United States government is *not* the United States. Government is not the society; it is not the community; it is not the state.

Government is *an* instrument through which the society or community can be said to act; it is *the* instrument through which the state acts. Thus, "Americans" look to government to do some things for them, but they look to churches and corporations to do things for them also. But when "The United States" acts, it acts through government; that is, "state action" is on the whole (this can get complicated) government action.

Now, in deploying all this I am, I believe, sticking very close to ordinary usage. (The only thing that might be controversial is my declining to fall in with a Marxist way of talking that identifies state with government, or with the coercive part of government. I see no reason to do that, especially since everything can be said in good English.) And, I trust, there is no misunderstanding of my thesis generated by confusion about terminology. The question is whether a community or society may use its political institutions in the realm of mind; whether government, by constitutional arrangement, can be given authority to act there.

9. *"democracy . . . is not an exception . . ."*

There is an odd tendency to confuse democracy with anarchy—some sort of view that one idea is as good as another; that all values are equal; that no one has a right to judge anyone else; that, in principle, anything goes, at least in the spiritual realm; and that it is an act of undemocratic tyranny to attempt to impart any order or form or structure into the life of the mind. Democracy, on this view, differs radically from all other forms of polity in precisely this—that it has no intellectual or spiritual foundation or presupposition or orthodoxy or doctrine, so that whatever government may do to create or encourage or protect or perpetuate the "democratic spirit" is seen as a fraudulent impertinence.

Unfortunately, or fortunately, this is simply a mistake. Democracy is not all that different. It is a form of government, perhaps a way of life, and it has a distinctive—non-anarchic—spirit. At its core is the significance it gives to universal (normal adult) participation in the political process, and the faith that all men can, if encouraged and given opportunity, develop the arts, the skills, the habits necessary for the life of responsible deliberation and decision-making. Democracy seeks to universalize the parliamentary state of mind. That takes some doing. Perhaps, in the end, it cannot really be done. Country after country grows weary of the attempt and gives up with the plea that the people lack the necessary understanding or education or habits or spirit. Democracy is not, I repeat, the natural or universal habit of the human mind; it is a cultivated spirit. And if we wish to maintain a democracy we cannot neglect its constant cultivation.

Democracy may, indeed, under some conditions, have a special

tenderness toward individualism and pluralism and a special
place in its heart for non-conformity and criticism. But even this,
and the tolerance, sympathy, and "openness" it requires, needs to
be diligently cultivated in the face of other tendencies—the tendencies of majoritarianism to self-righteous disregard of minority
interests, its brutal tendencies, and, as Burke says, its shameless-
ness.

10.   *. . . the free mind . . .*

There are justified anxieties at this point and I will be dealing
with aspects of the problem later, but let me say something in
passing. The worry is that the mind is in danger of being warped
out of its proper path, that its natural development is interfered
with by society (acting through government or otherwise) that
lays on its hampering cake of custom, strait-jacket, blinders, dis-
tortions, biases. There seems to be a contrast between the "natural" (universal? human?) mind and the cultured or convention-
alized mind and a suggestion that the latter is a corruption of the
former or at least a limitation of its real power. Left to its own
devices the mind would develop its innate powers naturally; but
we interfere and, like a Japanese gardener, dwarf it or shape it
strangely to our limited purposes.

It is not altogether easy to untangle this mess, to correct some
misconceptions while, at the same time, doing justice to the real
problem. It is quite all right to think of the infant mind in terms
of general human potentiality. But it is also the case that the
potentiality is realized only when it is fed by a culture. Thus, to
use a simple example, the potentiality for language is realized
only in the context of a linguistic community, and so one develops
as English (or Chinese or Russian) -speaking. It would be odd to
speak of this as putting one into an English strait-jacket or as
warping the mind from its natural language, or as limiting its
freedom. Sad, is it not, that one should have no choice of his
mother tongue. What an imposition; what a flouting of freedom!

But more than language is involved. By this same culturing
(choiceless) we take on our second nature—habitually driving on

the right side, fork-using, successively monogamous, hand-shaking, movie-going, Robin-hood loving, baseball watching. We are children of a particular culture, and we show it. Warped? Tricked? Imposed upon? What side of the street should we really drive on? You mean baseball is the wrong game?

Whatever fate metes out, it is really a fundamental error to treat the very incorporation of the given culture, the very condition of becoming a person, as itself a diminution of our freedom. But when all this is acknowledged, something still nags at us. Surely *some* culturing is warping, distorting, misshaping. Are there not sick cultures, engendering sick minds? Or sick episodes in the life of normal cultures? Can we not fall into the hands of a government that strains its energies to see that we believe what is false, hate what is different, distrust kindness, scorn intelligence and wisdom, enjoy cruelty, thirst for domination; that endlessly propagandizes, brain-washes, and practices the crudest arts of thought control?

Yes, we can. We have seen it happen. It is horrible and we must prevent it if we can. But how? Not, at any rate, by denying to society, state, or government any authority in the realm of mind. The free mind is not the product of even benign neglect.

11.   . . . *to lighten a burden of guilt and incoherence* . . .

This does not quite express the situation. I am writing this because I believe that the conception of government as an intruder in the realm of mind is a fundamental error; because I think it is an error that is widely shared and that does a great deal of harm. It makes it more difficult to understand what we should be doing. It has something to do, I believe, with the disarray, with the foolish excesses that seem to baffle us in such basic institutions as the school and the forum, with the strange incoherence of the Supreme Court—our most reflective political institution—as it grapples with problems of education and freedom of speech. I really think it would be better for us if we start out by making sense rather than by unfurling a foolish banner.

And, in my experience, we do begin with the banner. "Govern-

ment has no business in the realm of intellect and spirit" uttered
before students, or faculty colleagues, or even in the world beyond
the ivory tower, evokes a casual but deep "Amen!" To challenge
it suggests that one has forgotten Hitler or Joe McCarthy and has
never heard of the Civil Liberties Union. But my experience is
also that, when the long story has been unfolded, my hearers—if
any remain—are merely bored by the obvious insight, if anything
so banal can be called that, that of course government is legitimately concerned with the mind. There is some predictability in
initial response. Conservatives of the "all government is evil"
school resist the intrusion of government in this as in any domain.
Radicals and revolutionaries are suspicious; they can hardly condemn governmental involvement with the mind without breaking
with the traditions of Lenin, Stalin, Mao, or Castro; but they are
prepared to marshall some indignation against the wrong ("oppressive") government's mind-warping actions at which, naturally,
they are not surprised. Liberals are more fun since they (they! a
dozen stale varieties) tend to proclaim "hands-off the mind"
most fervently even while demanding that government do
something at once about ignorance, prejudice, television violence,
advertising, etc. In any case, the slogan does not survive reflection.
When I was a bit more naïve I was surprised to find that there
were some intellectuals prepared to oppose to the slogan "the
truth shall make you free" the insight "this error will keep you
safer." It does not surprise me now. I have heard too many arguments about how the truth is dangerous because it can be mis-
interpreted, can serve as an entering wedge, can be used by those
with bad motives, and is, in any case, unnecessary since we have
our instincts that steer us properly, ignoring whatever goes on in
the futile realm of abstractions. "Government has no authority
. . . etc." is the way to talk when it is necessary to talk at all; it
sounds good, free, American. And who will notice that it makes
no sense or, if noticed, really care? Just another example of how
theory can confuse honest simplicity.

Between banality and danger, between "don't belabor the obvious" and "let sleeping dogs lie" I find myself a bit unnerved.
It *is* so obvious. (But why, then, don't we believe it?) It is dan-

gerous. (But why, then, do we ignore it?) Why be a nag? So, I find myself rewriting and rewriting and, with each rewriting, pruning, eliminating, cutting down, until this opening section I once so lovingly deployed—the complex genesis of misunderstanding, the elaborate analysis of the spiritual nature of the human community, the careful dissection of authority and legitimacy—has reached its present shrunken, cryptic form.

I will let it stand, with mixed feelings. On the one hand, it is a fundamental part of the argument; on the other hand, it is perhaps the least convincing part of the argument. The argument about the legitimacy of government's concern with the mind has two parts. The first I have just gone through. The second, that follows, simply piles up so many examples of virtually undeniable legitimate involvement that it becomes ludicrous to raise the question, "but is all this legitimate?" The counter-examples swamp the general principle. And they do it so irresistibly that, from a rhetorical point of view, the first part is unnecessary.

Why, then, do I keep it? Because, whatever its rhetorical force, it is, I think, important and necessary. It remains, to my mind, "the first part."

CHAPTER 2

12. *Awareness failure*

"False" and "invalid" are probably the most familiar categories of cognitive failure, bearing respectively on belief or statement or sentence or proposition and on inference or argument. The analysis of perception may make use of "non-veridical" or "illusory" in connection with flawed perception of "what is there"— as in hallucinations. "Errors" in belief, inference, and perception are relatively simple or clear-cut ways of being mistaken as compared, for example, with "not paying sufficient attention to" or "not realizing what was happening" or "failing to see the significance of" or "failing to notice." The latter are also regarded, properly, as cognitive failures, but they are strangely complex. "Failure to notice," for example, is not just a case of not noticing

something that is there. "Not noticing" is, by itself, not necessarily a mistake. It is a mistake when it is the not noticing of
something that turns out to be important and that, had it been
taken proper notice of, might have enabled us to avoid some evil.
We are likely, after the event, to rue our failure to have grasped
or seen the significance of something.

Failure to grasp the significance of something is a fascinating
kind of mistake although, in some cases, we may be inclined to
challenge its characterization as a mistake. Idly to watch the
smoke curling up and not to grasp that it signifies "fire" would
probably be admitted to be an awareness failure. But to count the
dead and, in grief, to fail to grasp the significance of a victory may
be a failure of awareness that some might deny to be a failure at
all. The poignancy of the immediate may swamp the awareness
of significance. We may be blinded by tears—or by fears or by ela-
tion—to the significance of what is happening.

As against "failure" one may wish to speak of "levels" of awareness, leaving aside, initially at least, invidious distinctions. But
"level" suggests higher and lower, and the deficiency of the lower
inevitably comes to mind; although, as I have suggested, this may
be challenged. Let me present my favorite example.

In the *Iliad* Homer moves between three levels of awareness,
that I will call the plebeian, the heroic, and the Olympian. The
plebeian awareness is expressed by Thersites, for whom (and for
the multitudes for whom he speaks) the war is a protracted and
uncomfortable confusion. Hurry up, wait, advance, retreat, dig,
scrounge, scratch, shiver, bleed, and, sometimes, die. For what?
Occasionally a glimpse of strutting leaders, gossip about rivalries
for glory, some regimental solidarity, garbled messages from
Olympus. At the slightest suggestion, they are ready to sail for
home. Thersites expresses the G.I. point of view, from which the
whole expedition is pointless misery. Why are we here? The
misery is real; everything else is fraudulent nonsense.

The heroic awareness, dominating the famous warriors, the
glory seekers, is also fairly simple. More removed from misery and
more aware of the mission, of strategy and tactics, the hero is
chiefly aware of what things signify for his career, his war record,

his fame. He is aware of the war as the scene of glory, and Homer feeds our interest in all this with the exploits of Hector and Achilles and the rest.

The third level takes us away from the mud and the glory on the Trojan field to the position from which the real political issues can be seen. We cannot understand the struggle until we see the alliance between Aphrodite and Ares (love and war) against Hera and Athena (home and city) until we understand the Olympian or higher politics behind the struggle between Greece and Troy.

Can we not say that the failure to see the big picture—to eke out what is immediately before us with an awareness of its import—is, in fact, a failure of awareness—not just a false belief, not just an invalid inference but something worse, something deeper, something not so easily correctable. We are, it can almost be said, in an age of Thersitean awareness, undeniably aware of some undeniably real aspects of the life in any expedition for a cause. But the Thersitean awareness misses the significant point. Not the intermediate point of the generals or of the glory-seeking careerist but the ultimate point of politics—freedom, dignity, justice. To lose the awareness of that to the sobs and itches of immediacy is to live with the deepest of awareness flaws, the blindness of the lower echelon to which everything "higher" is unreal.

13.   . . . *the strange world of videality*

Plato's allegory of the cave is a dramatic way of presenting the contrast between appearance and reality, between sensing and understanding, and it anticipates the conception of images projected on a screen. But the screen has now grown to such massive proportions, it is the focus of so much of our attention, and the politics of the projection of images has become so crucial, that I suggest the notion of "videality" as stressing the edited quality of what the screen presents, as against the raw, unexpurgated flow of images that still, in Plato's terms, makes up the world of appearance. Thus we have the general world of appearance, the world of screened appearance, and, perhaps, the real world or the world understood. The additional layer adds some interesting implications that it is fun to explore. A world first taped and canned, played, delayed, replayed. Imagine falling farther and farther behind. Imagine . . . the possibilities are endless. A new world with its own rules and ethos. For more and more people, perhaps, the only—the "real"—world. In this context, "who edits?" is really an earth-shaking (or world-creating) question. Will we end up living in the "official" version in a single-channel world?

14. *Don't shoot the messenger!*

The claim of the media to the blamelessness of the messenger who is merely bringing the news, not making it, has really lost much of its point. The decision to report something, to make "news" out of an event, is more than bringing the news as it happened; it is deciding what should be called to our attention. Imagine a messenger with a hundred "messages" of which he may choose the one to deliver. May he not be asked why he delivered *that* message and not another? And even blamed if he delivers a trivial one instead of an important one? And if it turns out that the messenger has a delivery policy of his own, based on his judgment of social importance? The image of the poor, hapless messenger begins to fade. Of course, what he reported "happened"—let us suppose *really* happened, unrigged, unstaged. But so did a hundred, a thousand, other things. "Delivering" shrivels in its significance beside selecting, repeating, dwelling on, following up, harping on. The press may not make the news, but it certainly is responsible for its make-up. That is why, I think, popular sentiment refuses, and rightly, to be altogether taken in by the pose of mere "reporter," by the plea of "but that's what happened," by the guise of innocent messenger. It realizes that, in some real sense, the press is responsible for the news.

It should also be understood that the selective spotlight does more than illuminate. It distorts and misleads. It brings a patch of world under scrutiny with an intensity that may obscure the relation of the patch to the rest of the fabric. Or to change the metaphor, it is as if we glance at the world and suddenly, without really being aware of it, are seeing a microscopic version of a small piece of it. This produces some serious misconceptions about the way things are.

The fact is, when we zero in on a small segment, turn on the spotlight, pull out the microscope, examine with intensity, we run the risk of losing our sense of proportion or balance. We can focus so intensively on political prisoners in Cuba, let us say, that we may quite forget that our intense awareness of *that* does not imply that a similar situation may not exist elsewhere. But if we respond to what the spotlight illuminates—wherever it happens to point—our action may be ludicrously unbalanced. In this obvious sense, our focused awareness may mislead us not through ignorance but through disproportion.

15. *"Health, Education, and Welfare"*

The federal agency that deals with these matters is, without additional authorization, already involved in the politics of awareness. In enforcing the Fourteenth Amendment's requirement of "equal protection" it tries, primarily in the field of education, to prevent discrimination against racial minorities (and against women). There is a tradition that, by virtue of the Fourteenth Amendment, the Constitution is "color blind." That seemed to mean that official action could take no notice of color, that we were to be unaware of it. Thus, color or ethnic or racial identification on application forms was regarded as improper. To the question, "How many blacks are there in your school (or office)?" the proper answer was to be "How should I know? I don't notice such things." People were to be treated according to merit and, since color was irrelevant to *that,* it was not to be taken into account or noticed. Color was considered an improper category for the sorting of people.

This of course meant (how old-fashioned all this sounds!) that quotas were out, since to have quotas required keeping track of color (or race, or sex), of noticing systematically what was to be beneath notice. And quotas are, presumably, still "out." Although—and now we get a bit confused—how can we tell if our policies do not reflect discrimination if we do not check to see how the ethnic proportions are working out? So, to the surprise and indignation of those who had practiced color-blindness, we were told to start counting by race or color or whatever. Just to see how things stood. And then, if it turned out that our proportion of X's was less than the "proper share" (lots of interesting assumptions here) we were to demonstrate our lack of discrimination by bringing ourselves up to snuff. Not with quotas, of course, but by setting goals and time-tables and engaging in affirmative actions that, taken together, can be distinguished from quotas by a heroic effort.

This, as anyone knows, is not a simple problem. The politics of classificatory categories can get fairly elaborate. Do you include the Portuguese with the Spanish? Must you keep track of homosexuals? Can you notice race and sex for a little while only and return to not noticing when everything has been "rectified" according to some theory about the way things should be (merit? racial parity?)?

The Department of H.E.W. has been the federal administrative agency most actively involved in awareness politics. Its record does not fill everyone with confidence in the ability of government to cover itself with glory in the directing of awareness even though government may, at times, enrich our lives, as by adding "chairperson," by legislative fiat, to the armory of human dignity.

Freeing the question from the merits of H.E.W. or the sensitivity of government in general I would stress the importance of deliberate concern with the structure of noticing, with cognitive categories. There are some important consequences attending the use of "he," "she," and "it." When (if ever) should "black," "Jew," "felon," "female," be made obligatory matters of notice—stamped on identity cards, stamped into the awareness of schoolchildren, stressed in news accounts? What must we notice? What must we be required to ignore?

16. *The autonomy of the profession* . . .

That the basis for the claim to professional autonomy is the relinquishing of the rights of partisanship in the pursuit of the profession cannot be emphasized too much. This holds of the professions generally, although I am here concerned only with the "cognitive professions"—primarily teaching, research, journalism, and psychotherapy. For research and teaching the self-denying ordinance is embodied in the charter of academic freedom. For the cognitive professions that normally work outside the framework of public office or public institutions the situation is not so clear, although the professions, with authority over licensing and discipline, claim to live by and enforce a code of professional behavior. That code, with varying detail, invariably stresses the obligations of the role and the commitment of the practitioner to the demands of the role in the face of his private predilections, biases, or "politics." Some degree of spiritual celibacy is the price paid for the privilege of being trusted, for autonomy. The more it becomes apparent that a profession no longer conceives of itself in this way the more does it make sense to substitute external constraints for the self-restraint that has been disavowed.

17. *Dionysus*

In the *Bacchae,* Euripides depicts the arrival of Dionysus at the head of his frenzied and intoxicated band assaulting the neat Olympian rational world. It is a great drama, and many of us had the privilege of ringside seats at a return engagement in the mid-sixties. The tumult has subsided, but something remains of the strange alliance between drugs and revolution—the profitable radical exploitation of Lenny Bruce. There is nothing terribly original about the American version. It simply borrows the basic culture pattern of the bohemian brown-shirts of the Hitlerian Storm Troops. There is something easier than religion to serve up as the opiate of the masses.

But less "political" gurus have also made their pitch—some now publicly repentant—about freeing consciousness from its cage by popping this or that, and a vast activist claque is snarling and whining about the decriminalization of its favorite pacifier.

China, some time ago, in what is remembered as a heroic cultural episode, fought to free itself of the enlightenment of opium. We think of the imperialists who spread the habit as the villains of the piece. The cast of heroes and villains is a bit confused now, but we can all await, breathlessly, the outcome of the attempt to improve the world by cutting through the alcoholic haze with the bright blade of hashish.

18. *the university . . . the seat of knowledge.*

Everyone has always said that "knowledge is power," but this has seemed to be a conventionally pious remark, a kind of joshing admonition to students to study harder. And perhaps, when there was less of it around, it appeared less powerful than gold or armies. But that was long ago. That knowledge is power, or the indispensable basis of power, is today simply a flat statement of what is the case. Modern technology, industry, communication, computation is utterly dependent upon a highly sophisticated development of its cognitive basis. Without modern knowledge there is, in the modern world, no real power.

To see the "university" as, so to speak, the powerhouse, the citadel whose time has come, is really not a wild flight of the imagination. Research is crucial. It is done in communities that cluster into a city. And the city has a governmental structure for dealing with both its foreign relations and its internal affairs. The university is not identical with its government, of course, but the operation and nature of its government have some bearing on the quality of its life.

19. *"an attractive way of life . . ."*

This is a troublesome problem for dictatorships—especially those with a heavy ideological basis. The professional intellectual, scientific and otherwise, is likely to share professional interests without regard for political lines and would be handicapped severely if cut off from all contact with his international community. Some freedom from the constraints of the local ideology is an inevitable part of the productive way of life of the scientific community. Government may, if it is ideologically zealous, put its commissars or priests in charge and insist on orthodox science or, needing results, may, in a "pragmatic" mood ease up on ideology and give science some of the "freedom" it finds both necessary and attractive. This freedom to take the demands of orthodoxy lightly has a threatening aspect and gives rise to the familiar conflict between "red" and "expert," between "priest" and "pragmatist," that, for example, may be part of the so-called Cultural Revolution in China. The zealots want to bring the cognitive arts to heel; the cognitive agents want to get on with the job—that requires among other things, a life lived in the style of the academy, not on a farm.

20. *Contributions . . . strengthen the claims to support.*

There are peculiar controversies in the university, from time to time, over what it supports, aids, or cooperates with. Agriculture, industry, business, the military—these and other interests have made their treaties or deals, and the university shows the effects in its schools of agriculture or engineering, or law or business administration, in institutes of labor relations or international affairs, in doctor or dentist or athlete or officer training programs. So varied are its activities and the interests it serves that forgetting perhaps, what "university" means, we have come to speak of "multiversity."

To the monastic type, the university is conceived along more austere lines; it is a place of research that is pure or, if there is any doubt about it, useless. Anything else is a compromise with worldly ambition and a betrayal of the pursuit of truth for its own sake. As a purist, he gazes with disdain at the burgeoning barbarous precincts and wishes they all would go away.

Others would wish to banish more selectively, to get rid of or fend off activities that they deem incompatible with a conception of the university a bit more hospitable to practical concerns. But the criteria may not be very clear. Thus: medical school? OK. Law school? OK. Business school? Well . . . . Or it may be deemed proper to include a school of dramatic art but not a school of physical education, sensing somehow that a professional actor "belongs" but a professional basketball player does not.

And then, of course, there are the "politicals" who may hold that helping consumers is good but that helping corporations is evil; or that the university should help employees, not employers. Or, of course, that lending the mind to the purposes of defense—the military—is a violation of cognitive morality; although why defense is not as legitimate a public purpose as anything else is not really explained. Perhaps in *good* countries . . .

Thus, the shape of the cognitive city is a matter of some internal dispute. There are many warring conceptions, and I point to this as an example of the pervasiveness of the politics of cognition.

21.   . . . *without endangering the supply*

There is a paper curtain interposed between the appropriating legislature and the particular research project that usually shields some exotic research from the unsympathetic scrutiny of the practical taxpayer. But sometimes a politician looking for a soft target will use his investigatory powers to do a bit of egg-head baiting and loudly demand an explanation of why "the people" should finance inquiries into the history of the button, or the sex life of goldfish, or studies of unsuccessful novels (1900 to 1915). There is general embarrassment at such demagoguery—and it *is* usually demagoguery since anyone astute enough to know where the bodies are buried also knows *why*, but wants some cheap points. Good reasons can often be given, but the mood in which the question is asked is seldom the mood patient with long answers. The point is not that something fraudulent is going on, but that the rationale for a particular piece of research is found in a context that is the research community's way of life, and it may be difficult to explain at every point—especially since some things that are done may well be pointless. Block budgets, some internal discretion, and some decent obscurity help the institution. Also, a good football season.

22.   *formally forbidden areas of inquiry*

Apart from the conception of a sacred area, protected by taboo from profane inquiry—a conception that may still have some slight life—the notion of forbidden knowledge is related to knowledge that is dangerous if used or misused or that, by its very existence, tempts us into doing things we should not do. That we now know how to destroy human life on the planet is, of course, the obvious case of dangerous knowledge. Suppose, some years ago, aware of this possibility, the world, by treaty and with adequate inspection and enforcement had declared a ban or a moratorium on nuclear research. Would that have been an unreasonable thing to do? The usual fear, and it may be well founded, is that enforcement is impossible and that, in a competitive world, it is more dangerous for us *not* to have the knowledge, since they will. So cognitive disarmament tends to give way to the balance of cognitive terror.

An example of knowledge that may be less cataclysmic but that, it is feared, might further "bad" conduct is the knowledge of race-connected differences that, should such differences turn out to be significant, might encourage acceptance of "caste" or racist policy. This fear expresses itself in "no testing!" provisions in public schools, heedless of the plea that it is always better to know.

CHAPTER 3

23. . . . *minor and adult* . . .

I realize that, in what follows, I devote most of my attention to the teaching power as it comes to bear upon the minor and where it operates, when challenged, with the powers of compulsion. That is its most difficult and controversial sphere and the one in which the challenge to its legitimacy, when its claims are pushed to the limit, is most familiar. But I do not mean, by this emphasis, to decry the importance of the teaching power as it exercises itself in the world of adults where, as I suggest, the issue is not the scope and nature of compulsory schooling but rather the access to special opportunities. The competition for places in professional and graduate schools has become quite fierce and the politics of opportunity quite bitter. Traditional admission criteria—grades, scores, recommendations—are challenged as racially, or sexually, or economically biased and there is a push toward quotas and "affirmative" admission procedures that, in some cases, are being challenged as "reverse discrimination." An implicit assumption is that an "unbiased" distribution of professional schooling opportunity would produce professions that are both "representative" and maximally competent. But even if this were not the case some difficult questions would remain. For example, if the ideal criteria filled our medical schools entirely with white women, there would be legitimate political questions about departing from the "ideal result"—the "most skilled" medical profession—in order to provide for existence of some male and some nonwhite doctors for reasons of social policy. There is a growing body of literature on such fascinating questions and my bypassing them does not mean that I do not consider them important.

There are other questions as well as these more familiar ones. Most of our teaching energy is directed at the young and even our graduate and professional schools are inhabited largely by those within a decade of adolescence. But as we become affluent and live longer and entertain possibilities of mid-career vocational changes and become accustomed to early retirement, the redirection of the attention of the teaching power, in various ways, to the adult population becomes an interesting question of policy. Adult education may cease, someday, to be an oddity.

Nor do I deal here with the questions of "re-education," compulsory or voluntary, that really deserve some attention. Some regimes are determined to "re-educate" significant portions of their populations in concentration-campuses. Others limit compulsory re-education attempts to the rehabilitation of those in reformatories or to those whose mental health leaves something to be desired.

There are, in short, a great many aspects of the exercise of the teaching power of the state, broadly conceived, that I mention only in passing. I cannot really defend this as a complete account. My lame excuse is that "Government and the Mind" does not mean "Everything that should be said about Government and the Mind."

24. *Strong and weak versions (Sovereignty and Pluralism)*

The question is, "Who shapes the young?" And the initial response inevitably puts the family in the center of the picture. But
the family is not autonomous. It is constrained by the law of the
broader community. It may drift in the mainstream of religious
life or may be caught up more strenuously in a sectarian enclave;
it may dwell in social and political orthodoxy or stamp itself with
the deep mark of a dissenting or heretical movement. It brings up
its children not merely for family life but for something more.
But its control over the situation seems to have diminished drastically. The family may no longer control the significant educational environment of its children. The movies, the news and
magazine rack, radio, television, the neighborhood are competing
influences almost impossible to overcome. And the children are
yielded up earlier. The remorseless pushing back of the home-
leaving date—kindergarten, nursery school, pre-nursery school,
day-care center—interposes ever earlier between parent and child
another set of adults serving it is not always clear whom and
equipped with God knows what sort of weird notions. In relatively free situations the parent may stand bewildered, reduced to
insignificance by a seething Babel; in dictatorships he may be
rudely elbowed out of the picture, reduced to subservience by
children who have been taught to report on backward tendencies.
The condition of the family in the contemporary world does not
really permit us to rest with "the family" as a conclusive answer
to "Who shapes the young?" or even to "Who should?" Still, the
family, and its condition, must be kept in mind as we consider
the claim of the state to the teaching power in its strong or in its
weaker form.

I begin with a reminder of the stark simplicity of the strong
claim as it is asserted, for example, in states like the Soviet Union
or Cuba or China. Where the problem of education is seen in
terms of raising a new kind of person, of creating a fresh or correct consciousness free of the corruption of an old system, the
triumphant revolutionary power simply takes over the incubator

and rules it with a jealous eye. We hear little *there* of "private" schools, of alternative schools as counter-cultural enclaves, of old-style or deviant adult values interposed legitimately by parents between the child and the ministers of education. The teaching power, wielded exclusively by government, brooks no challenge (although it may sometimes condescend to explain to awed gapers from abroad, whose children are safe in alternative private schools back home, that someday in the happy future . . .).

The assertion of a direct governmental monopoly of the teaching power is hardly familiar on the American scene and is not advocated by left, right, or center. Compulsory schooling, yes; with some marginal dissent—generally not, it should be noted, by the "disadvantaged." There are disputes about substance and standards and duration. But there is general agreement that compulsory education in America does not mean compulsory public school education. Our educational establishment has three parts: a massive public system, nursery school through graduate or professional school; a large, almost massive parochial or church-related sector that, under our "no establishment" commitment, is broadly hospitable to almost all claimants, is without formal distinction between orthodoxy and dissent, and is largely, in principle, without public funds; and a large private sector, neither church-controlled nor public, also ranging from nursery school through the graduate and professional school. Clearly, the arrangement is not one in which the government's claim to authority to teach is asserted in a monopolistic form.

But even this proliferation of teaching institutions does not lay the possibly overriding claims of government to rest. We require "schooling" up to a certain age. We undertake to provide a place in a public school for everyone subject to compulsory attendance. The parent may decline the public-school option. Truancy is avoided by the provision of equivalent schooling, and equivalence requires governmental certification. The certification or accreditation of schools as equivalent may, in practice, be quite lax, but the principle of the authority of government to set standards is preserved. The parent is permitted to offer attendance at an accredited school as satisfying the compulsory schooling requirement. It is not quite the case that no school may exist without the permission of the government; it is the case, however, that no school can offer itself as satisfying the attendance requirement unless it has been authorized, by government accreditation, to do so. Something of the "strong" assertion of the teaching power is preserved in this form, even though there is no public-school monopoly.

The point of our arrangement lies in the choice and variety it offers. Since we are dealing with children or minors the choice is exercised on their behalf by parent or guardian. It may be an exaggeration to characterize as "choice" what may be so much a result of circumstance and family habit but, within limits, the choice is real. It may be based on anything from considerations of class size and curricular richness to pedagogic style and religious orientation. The value of the arrangement is not only that it does satisfy deep and differing convictions about how children should be educated but also that, by providing alternatives, it removes from the public school an unsupportable burden of controversy that, if there were no alternatives, would rage bitterly within it. And, for those who are neither substantively partisan nor impressed with the avoidance of trouble, a complex, varied system of schools can be seen as insurance against serious error, as a way to avoid putting all our educational eggs in one basket.

In spite of alternatives and variety there are, however, two points at which the authority of government asserts itself. First, there is the question of a governmentally insisted-on minimal core that all schools must provide; and, second, there is the question of governmentally declared and enforced limits to what may be taught in any school. There are some things that all schools must do; there are some things that some may do that others need not or may not; and there are some things that no school may do. In spite of efforts to restrict the teacher power to a weak form—in which government merely operates its public schools as one element among others—the authority of government inevitably asserts itself in guarding both the essential educational core and the bounds of educational legitimacy.

Consider, first, the problem of the essential core. Since this is normally a minimal requirement the argument for it is more likely to be based on the needs of the child, if he is to have his fair chance, than on what is essential to the society, although that may also be involved. Thus, a polity may have a mandatory universal language requirement: everyone, let us say, must be taught English as the primary language, whatever other language he may be taught; or, perhaps, everyone must be taught both English and French if the polity adopts a policy of cultural bilingualism. The core may include the elements of calculation or mathematics and some history and social studies deemed essential for life in the community. And we may, as John Stuart Mill suggested, require the cultivation of competence up to a certain level. All this seems unobjectionable enough. The public schools provide at least the minimal core and a non-public-school alternative must be certified as doing so as well. But there are objections. Some parents may consider some part of the core as not necessary at all and even as harmful, and some may consider that it does not include what is really essential. There may be some attempts to reshape the public requirements to these views before taking refuge in private or parochial schools. Thus, we have had battles over the inclusion in the public school of a religious element considered, by some, an essential part of any sound education. And, in the case of the Amish, the required level of education was regarded as producing a worldliness that threatened their way of life. The core included too much; it might enable a boy to leave the farm and the community. It may seem odd that the Supreme Court heeded this plea and acquiesced in the parental demand for a lower minimal core for their children to deprive them of mobility, but in any case it took an action by the Supreme Court to sanction this anomaly.

A core requirement the polity insists upon is something that all schools, public or private, must do. Where that is not considered enough, the parent may seek to supplement it in various ways; but where that is considered too much, the process of certification of alternative schools makes the requirement, in principle, inescapable. *What* shall be required is a part of the politics of education.

There are times when the intervention of the government with its insistence on what is required appears as a rescuing of the child from parental tyranny, from a pattern of hereditary ignorance, from a narrow, limiting irrationality, from a depressingly benighted pattern of culture—the parent grimly following the child to school with "I don't want him reading this. I don't want him hearing about that . . ." Or taking him out of public school to put him in his own safe special culture-preserving school to evade, if possible, even the minimal demands of the general culture. But soft! Do we believe in pluralism? In real differences, not merely trivial variations? Are we serious when we assert our belief in the value of *different* styles of life? How different? Can some differences be beyond the pale? The answer is, "Some differences are intolerable."

That is to say, the principle of pluralism, like the principle of toleration, has its limits. I cannot understand why this should be treated as a scandal. When we learn that more than a single way will serve, we open up to a range of acceptable alternatives. Not to everything. The prohibition of some "alternatives" is compatible with pluralism. If we give up the principle of a single religious establishment and recognize, pluralistically, Catholicism, Judaism, and Protestantism, it does not follow that we must also recognize atheism as an alternative on an equal footing. The principle of pluralism does not require universal hospitality. Some things may be out of bounds.

So, also, with educational pluralism seeking scope in our complex system of public, parochial, and private schools. I have already argued that a common core can be required. I now suggest that some things, some culture patterns, may be educationally out of bounds. I will not parade horrible possibilities. Let me refer, instead, to the growing political quandary over "school busing" and let me put it as a question.

What is the real issue over busing? Not a ride in a vehicle. Not even equal or "quality" education. The real issue is over the determination of government to interpose itself between the child and the parent who, in his bones, wants to bring up his child in his environment and his culture—white, middle class, ethnocen-

tric, color proud—the culture of his home, his neighborhood, his
mores, his traditions, his ethnic jokes and foibles—the way of his
fathers. What are we to say? "That's pluralism!" or "Not *that*
culture; its against public policy." We can ban it, let us say, in
the public schools. But suppose it moves massively to the private
school system. Will we still want to say, or be able to say, "That
culture pattern in the schools is beyond the bounds of American
pluralism?" If there is to be the exclusion of an educational pattern from the sphere of legitimacy—whether we take the above example or other examples—the only effective exclusionary agency
is government. It is the only possible guardian of the limits.

Thus, I believe it is apparent that the teaching power of the
state, even when it is asserted in its weak form; even when it
eschews the right of monopoly, retains, in the power to require
and in the power to exclude, something of the final power that,
ordinarily, we associate with the right of sovereignty.

25. . . . *beyond challenge* . . .

There are, of course, challenges to the public-school establishment
that direct themselves not so much to the legitimacy as to the desirability of the system. The hysterical critics who would abolish
the schools that they see as warehouses or prisons for the young
do not deserve notice. The faddish vogue of "de-schooling"
would, if anyone listened, merely doom the "disadvantaged" to
permanent hopelessness. The attack on the privileged claim of
the public school to tax support expressed in the "voucher" movement is more interesting and does raise a significant issue about
fairness in support of educational variety. It also offers the benefit
of genuine competition to a complacent institution. And, in various ways, the growth of private schools as an escape from the
turmoil and inadequacy of the public school, challenges the domi-
nant status of the public school. In many ways the public school
comes under increasing criticism. Most of the criticism is well de-
served, but it should not obscure the fact that the public school
is an amazing institution, relatively recent in its massiveness, bril-

liantly successful in giving a concrete form to the ideal of free (relatively) education for all as far as ability, energy, character, and luck will carry. In the midst of our frustration over its disarray and its obvious shortcomings we should pause, from time to time, to acknowledge the incredible success with which it performs its impossible task.

26. . . . *sanctions* . . .

The ultimate sanction of the teaching power, expulsion, is really too drastic to be used freely. It can relieve the school of the burden of dealing with a troublesome and intractable youngster, but it dooms him to a life with limitations of which he is not yet really aware. The school, it should be noted, really loses nothing by expelling the hard case; it puts up with him for his own good. If we didn't care about him the school could restore order with relative ease. Expulsion is like spiritual abortion.

But the consequences of the exercise of teaching power authority are drastic in other than disciplinary contexts. Tracking and guiding into roles may have decisive and permanent effects. The certification of competence or fitness for higher educational opportunities is the modern substitute for the accident of birth as the determiner of the quality of life.

It is, I believe, no exaggeration to say that the consequences of the exercise of the teaching power outweigh the sanctions of the judicial power.

27. . . . *wanting to teach* . . .

What Plato says about politics seems to me at times—although I hesitate to say so—applicable to teaching as well: that eagerness for the role is more likely to be a sign of unfitness than of fitness. (My hesitation is due to the fact that this runs counter to the normal pieties of my profession and I am growing suspicious of heresies even when they are my own.) I have in mind less the teacher of a particular craft or art who, as part of his practice, initiates others into its mysteries than the person less concerned to teach something in particular than just "to teach" or to teach "students, not subjects." That suggests to me a dangerous disposition to impose oneself upon others, an eagerness to shape the malleable, a confident egoism, far removed from the spiritual condition of the true teacher. Preacher, perhaps, but that is quite another thing.

Nor do I consider that thus wanting to teach creates in any remote way the right to do so. Adults, of course, are presumably able to take care of themselves. But children are not fair game for spiritual pitchmen. It is odd that anyone should think otherwise; to think that he has a right to try to shape them or influence them or awaken them or lead them anywhere. Unless he is a parent exercising responsibility for his own children. Or unless he is appointed to the office of teacher by the proper authorities. The self-appointed teacher of the young is almost always a menace to everyone.

28.   . . . *curricular responsibilities* . . .

That the school cannot properly delegate or abdicate its curricular responsibilities should be obvious, but apparently this is not always the case. We have traditionally been alert to pressure from this or that interest group seeking curricular influence, and, when it works its way through school boards and educational authority generally it may be said to be a part of curricular politics and not necessarily illegitimate or, as going beyond proper politics, a case of educational irresponsibility. The irresponsibility most pervasive recently has been in connection with the demand for student determination of the curriculum—either individually, as in the abolition of requirements so that each student can elect what to study; or collectively, as when student representatives claim or are granted an equal or even preponderant voice in the establishment of educational policy. In either case to turn the shaping of a student's education over to the student himself or to his peers is so deep a betrayal of the student's interests and so complete an abdication of educational responsibility as to leave one speechless.

The smug, uncomprehending, "democratic" spirit in which this is done is, to my mind, more unbearable than the cynical indifference that simply bows to student pressure for the sake of peace and quiet.

29. *. . . denying him his proper liberty*

Defenders of the liberty and the "rights" of children spring up everywhere. It begins with the clear case of the battered, mistreated, or neglected child, surely an object of proper concern, and goes on from there. Soon we hear of ministers who offer themselves as negotiators between parent and child who have "differences" over questions of life-style; we hear of schools or teachers arranging for services for minors, bypassing parental knowledge or consent; of legal organizations for the defense of children's rights. There is a lovely aura of goodness about all this—the gallant rescue and defense of the weak and helpless. And yet . . . First, in spite of everything we say about the family, we systematically weaken its essential structure by diminishing the authority of parent over child. Second, we systematically weaken the tutelary power of public authority—as, for example, in the really ridiculous flag-salute opinion in the Barnette case and the more recent monstrosity, the Tinker case, that denied school authorities the power to ban a peaceful political demonstration in the classroom.

So we rescue the children. From their parents; from the state. Then what? Then we complain about alienation and pointlessness. We visit China and swoon in ecstasy at the sight of regiments of children pledging undying allegiance to their leader in schools from which no Supreme Court has banished a flag salute as an invasion of the rights of children.

There are other ways of battering children than by mistreating them physically—intolerable and inexcusable as that is. We can treat them as if they were adults and pretend that since "persons" have rights they all have the same rights. To treat those below the age of consent as if they have the rights of a consent-based status is not merely silly. It is suicidal.

CHAPTER 4

30.   *. . . the harm done by discrediting authority . . .*

This seems somehow to be a rather shoddy point. It is so easy to
see the angels as on the side of exposure, as behind the revelation
of the discreditable, of the feet of clay, of the moral slip, of the
private vulgarity or indiscretion. If that is what our leaders are
really like, surely it is better to know. How can we defend illusions? How can we justify unmerited respect? Is it not better to
have a crisis of confidence than to risk misplaced confidence?
Doesn't Watergate prove . . . ? The virtuous drift of all this is
clearly against the status of seditious libel as, so to speak, a legitimate crime. In oppressive regimes it is simply another aspect or
tool of oppression; in free polities it has no place at all. I should
probably let it go at that, especially since nothing in the general
position I am delineating requires any tenderness toward *lèse
majesté* and its ilk. But let me at least put the question in a dispassionate form. Is it ever reasonable or legitimate to hold that a
public functionary (or a public agency) while in the peformance
of function is to be exempted from or protected against certain
forms of criticism? The argument is that the performance of the
function is aided by, or even requires, protection against verbal
assault. Applied to government generally and concerned to maintain respect or confidence as a condition of effective functioning,
we have the traditional seditious libel. Applied more selectively
to a particular institution whose functioning is impeded by verbal
interference we have something more closely resembling "contempt." Perhaps it can be said of seditious libel that it is a general
extension of some aspects of "contempt of court" to the government establishment or, perhaps, to its chiefs. A utilitarian may
well find, under some circumstances, that protection of government against the weakening of its repute, the protection of the
community against loss of confidence, is called for by the calculation of benefits.

31.   . . . *verbal taboo* . . . *a neglected area* . . .

There is, of course, a large and notoriously confusing judicial
literature about obscenity as well as a popular literature that lives
off the judicial confusion and pushes the defense of obscenity as
a mode of free speech, from a civil liberties perspective. We have
heard a great deal about "but is it a work of art?" or "redeeming
social value" or "prurient interest" or "community [which? how
determined?] standards"—but on the whole (and could one expect
anything else?) the intellectual tone is amused or contemptuous
and the theoretical defense of attempts to control "offensive" language, whether pornographic or racist or sexist, is rather thin. As
usual, the decent instincts of the ordinary person are unprotected
by a corps of active intellectual skirmishers against the inroads of
snipers. I am unaware, although I may have missed it, of any serious psychological or anthropological analysis of the universal
phenomenon of the verbal taboo, quite apart from any concern
with current obscenity issues. And, on the latter questions, the
current rise of sensitivity of ethnic minorities to verbal subtleties
and the awareness of women's movements of the antifeminine
character of pornography may be bringing some new allies to the
dogged, embattled "puritan" spoilsport who has been holding the
fort alone for so long. There is, I believe, a perfectly respectable
case for the enforcement of the linguistic taboos that compel obscenity to perform its function in time-honored surreptitiousness.

32.   . . . *some supposed natural right of free expression* . . .

The pious misstatement of the spirit of the founding fathers and
the First Amendment is a ritual part of our approach to the discussion of freedom of speech. "They" are said to have expressed
in that amendment their devotion to freedom of expression as a
natural right beyond the control of government. This is, of
course, utter nonsense. Free expression as a natural right does
not remotely enter the picture. The First Amendment simply
nails down the understanding that the Federal Government (Con-

gress, more particularly) is not to interfere with the state regulation of speech, press, religion, assembly. Hands off states' rights, not natural rights! As everyone knows, or should know since John Marshall and the Supreme Court made it clear, the Bill of Rights was not intended as a limitation on the governments of the states, that were free, even, to establish religions if they saw fit and to regulate speech in various ways.

As everyone also knows, the defense of freedom of speech in our country in those days was a defense of the rights of Englishmen, not of the rights of man or of natural rights. In fact, it is striking that none of the major theorists in the whole tradition of natural rights lists anything like a right of free expression as part of natural right. The claim that freedom of speech is a time-honored natural right is a fraudulent claim with a forged pedigree. It is a short-cut to misunderstanding.

33. . . . *the hazards of advising* . . .

I would love to see a treatment of the institution of "adviser" from Delphi through Santa Monica, from the reader of auspices to Miss Lonely Hearts. Every society has and needs institutions through which persons—public or private—can supplement their own resources of understanding and decision. Advice comes in many forms, sometimes disguised as something else. The wielder of advice can have, obviously, great power and the institutionalized office of adviser has been carefully hedged. Great struggles have culminated in the establishment of the right to advise and the obligation to entertain advice. An interesting tension exists between the tendency to hold the adviser responsible for the consequences of the advised action and the insistence that, since the decision is the act of the responsible agent, the adviser be protected from blame for the outcome. We will soon resolve this tension, I suppose, by the development of advisers' malpractice insurance.

34. . . . *prior restraint* . . .

There is, I think, too much emphasis in our tradition on "no prior restraint." It is important and has, historically, played its part in the erosion of censorial offices. But it can be misleading. It obviously does not follow that what is not subject to prior restraint is therefore "permitted" and, accordingly, free from legitimate subsequent punishment. But I find that that is a not uncommon view. "Not subject to prior restraint" *is* confused with "permitted," and subsequent punishment, if it is proposed, is treated as if it were a dirty trick, a case of misleading entrapment. "No prior restraint" should be read as deferring the question of the propriety of an act of speech, not as settling it. In short, the emphasis on prior restraint tends to confuse the status of subsequent punishment as the alternative mode of regulation.

It is, additionally, foolish to treat "no prior restraint" as an invariable universal rule. Suppose the date of "D Day" is leaked or stolen and suppose a newspaper does propose to publish its scoop. Would anyone seriously argue that "prior restraint" is impossible, that the country must suffer the damage and only *subsequently* can punish the publisher of protected secrets? There may be dispute about whether publication would cause irreparable harm and *that* may be a question for judicial determination. But to hold that we must be content with subsequent punishment because we cannot ever prevent irreparable damage by use of prior restraint is to hold the kind of nonsense that only discredits the claim to proper freedom.

35. . . . *a failure in the morality of discourse*

I find students, especially those who are "concerned" about public issues, surprised and even offended at the suggestion that there is something wrong with putting pressure on someone who expresses a deviant political view by boycotting his business. Those with longer memories will remember how it was once thought improper to attempt to blacklist radicals, to get at those who were expressing unpopular views by cutting them off at the checkbook. This seemed a clear violation of freedom of speech—not to answer in the forum but to make the speaker or writer answer "in another place." It was, I believe, wrong then; it is wrong now. That a small-business man or a tradesman must, if he wishes to express his views in a letter to the editor, resort to anonymity because

otherwise the reward for his speaking out is that virtuous enthu-
siasts will try to drive him out of business, is a sad case of insensitivity to the needs of a free society.

36.   . . . *the schoolhouse gate* . . .

I find it difficult to express adequately how unwarranted I find
the sanctimonious waving of the banner of the First Amendment
as the Court blunders across the threshold of the school. How
anyone contemplating the routine exercise of the teaching power
in the school can still conceive that the principles that normally
govern the public forum apply there as well . . .

I suppose one should consider that the court is expressing the
view that the forum is ubiquitous and that, therefore, its rights
come to life whenever their exercise does not interfere with the
primary purpose of the situation. In my brief posing of the issue
between the "ubiquitous" and "adequate" conceptions of the
forum I leave open the question of which view makes more sense,
and I can, technically, live with either view. But I find myself
hardening, personally, against "ubiquity." Important as communication or the forum may be, it is not *that* important. It can be
excluded from areas dedicated to other purposes; it cannot sneak
in everywhere. And I'm not sure that everything else should always bear the burden of showing that the pursuit of its purposes
requires the exclusion of forum.

37.   . . . *a flood of information* . . .

The controversy over the power of congressional investigating
committees has subsided a bit. Senator Joseph McCarthy was
fond of quoting Woodrow Wilson who, in his book on congressional government, suggested that the informing function of Congress was more important even than its legislative function. The
classical view of the powers of Congress to "investigate" is that it
is implied by and is in the service of its power to legislate. That
is, the request for information rested on a need to know in order
to perform the legislative function; it had to show some relevance

to a legislative purpose. Using congressional power to "expose" merely to enlighten the public stood, and probably still stands, on rather shaky constitutional ground. The generation that grew up fighting the investigations of the Senate Internal Security Committee and HUAC (if you don't remember what this stands for, I won't remind you) might find if it raised, by ancient reflex, the question of the congressional or legislative right to expose and inform, that hardly anyone seems to understand or to care. "The informing function may be more important . . ."

38. . . . *the right to know* . . .

It should be noted that the practice of torture, where it is not simply indulged in as a pleasure or inflicted as a form of punishment, claims as its justification a right to know based on a need to know. He may know where the bomb is hidden; he may know where the terrorist headquarters is located; or where the kidnappers are holding the baby; or where the assassin lurks awaiting the revered leader. He is asked. He refuses to tell. He may even claim a right not to talk, not to incriminate himself, not to betray his comrades. We may, under a range of normal circumstances, even respect that right. But it takes little imagination to see circumstances in which anyone would regard torture as a justified way of getting some urgently needed information. When I say that the right to know does not override every other human value I must add that the right to withhold information does not override every other human value either. I believe in and support legal restraints that put torture out of bounds as a means of extracting information; I support, as well, international conventions that protect prisoners of war from being subjected to torture. This is an extremely unpleasant subject. It is unpleasant because, when we have said all this about making it illegal, there is hardly a person left to throw the first stone, who does not know in his heart that under the proper circumstances he would twist the kidnapper's arm . . .

It is, by the way, perfectly conceivable that the rules of war be revised to provide that a prisoner of war should be required to tell everything he knows—not merely the hallowed "name, rank, and serial number." And it is even conceivable that a system of criminal justice would be based on a presumption of guilt (if charged), a denial of a right not to incriminate oneself, and a requirement of full disclosure subject to very heavy punishment for refusal to talk. That, of course, is not our system and we would, presumably, in no way tolerate even a heavy sentence as a way of loosening tongues.

39. . . . *the effects of institutional arrangements* . . .

It is surprising how much the quality of discourse and, accordingly, of the mind, is affected by such things as the existence of a majority-rule (51%) provision as against more consensual requirements (⅔ or ¾ or even strong-interest veto power). In an assembly that can act by simple majority there is a tendency to ignore the most adamant opposition, to appeal to those closest, to sharpen opposition, to enact too much in the face of significant opposition. In contrast, the requirement of virtual consensus requires, if we are to act at all, that we communicate with the *deepest* opponent, that we attempt to understand and come to terms with him instead of treating him as the enemy beyond the pale. The frame of mind, the mode of discourse, the effort at understanding needed for consensual action (as I call everything from the requirement of unanimity or "significant" interest veto to something like ⅔) is quite different from the mood of simple majoritarianism with its built-in indifference to minorities.

Majoritarianism rests, presumably, on the obviousness of the one-man–one-vote principle and it is sometimes assumed that that is a requirement of reason or of fairness or of justice and that anyone, not concerned with his own special advantage, would naturally accept and even insist on "one-man–one-vote." That, I believe, is not the case. Knowing nothing of my own special status (behind "the veil of ignorance" as is now said) I find it more reasonable to accept, let us say, a two-thirds requirement for action, with its assurance that nothing will be done against the wishes of a large minority, and with the moral and intellectual habits it requires, than to opt for rule by 51 per cent as expressing one-man–one-vote. One would ideally like unanimity as a requirement, giving each member a veto power, but since it is recognized that that would destroy effectiveness, we move reluctantly away from unanimity. But Calhoun's principle of "concurrent majority" of, in effect, "significant interest" veto power, is more attractive, abstractly, even though, in weighing the minority voice so heavily, it violates one-man–one-vote. "Justice" does *not* mandate one-man–one-vote. Nor is it an arrangement that does much for the quality of communication and the deepening of understanding.

In America, fortunately, we are not too tied to simple majoritarianism. We still tolerate unanimity requirements for juries (1 vote outweighs 11?), 76 per cent requirements for constitutional amendments, 67 per cent requirements to act in the face of a veto and sometimes for local bond issues, etc. These departures from "51 per cent takes all" seem to me to be rather wise, civilizing provisions and save us, to some degree, from the folly of a Jacobin majoritarianism. I point out, by the way, that a generous provision for minority veto is an alternative to the resort to civil disobedience justified by majoritarian insensitivity. I offer this as an example of how an institutional arrangement affects the quality of the life of the mind in the forum.